GSM Made SIMple

Written by George Lamb
Illustrated by Yani Batteau

Cordero Consulting, Inc.
2793 Thornbriar Road, Atlanta, Georgia 30340
(404) 713-2919

First Edition Third Printing

Printed and Bound by Regal Printing, Atlanta, GA

ISBN 0-966-57520-2 29.00

I would like to thank the members of the GSM
community for their help in making this book possible.

This book is dedicated to Bonnie, my wife and best friend.

- George Lamb

Introduction

Human beings are basically a very clever lot. We do, from time to time however, shy away from a topic because it seems like it is just too hard to learn. Mind you, some things are more trouble to learn than they are worth; I still cannot program my VCR. However, many topics simply get more interesting the more you learn about them.

The introduction of advanced digital mobile telephone networks brings with it a wealth of benefits for mobile users in North America. It also entails the seemingly insurmountable task of learning how this complicated mass of bits and bytes and network components all work together to make a phone call.

And then, just to make it a bit more challenging, there are the acronyms. Try this sentence on for size: "Before we can activate a POP OTA, we first make sure that both the IMSI and the Ki are in the AUC of the HLR".

This is where I come in. I am a telecommunications engineer. For the past fourteen years I have worked with mobile telephone systems throughout the world. While I have to admit that I do get excited about some pretty weird stuff, like finally understanding a particular signaling path, I do not wear a pocket protector.

In order to understand the complex interworkings of mobile telephone systems, I try to put them into simpler terms that my brain can absorb. A few years ago I realized that there are many people in the business in need of as much (or more) understanding of these systems as I am; but, like me, they need easier concepts in order to learn how it works. I also found out that the manner in which I simplified things for myself helped me to explain things to others. Now I spend most of my time teaching advanced technical concepts in plain English.

What I realized, is that in most cases, technology, like that used in the Global System for Mobile Communications (GSM) systems, is not magic at all; as a matter of fact you do not even have to be a rocket scientist to understand most of it. You just have to find some one who can explain it to you.

Thus the concept for this book. A simple explanation of a very interesting topic, which I hope will prove that.......

You Don't Have To Be An Engineer To Understand GSM!

A Note To Engineers (and other serious technical types).

If you purchased this book to pick up some tidbits about the MAP layer of the SS-7 protocol, or any other deep technical data information, TAKE IT BACK! You won't find that kind of information here.

As a matter of fact, if you do read this book, you will find several places where I have greatly simplified the process to make it easier to understand. For example: we know that when information about a particular user is passed, the GSM network uses a TIMSI rather than the actual IMSI. Unfortunately, discussion about TIMSI's and pseudo MSISDN's really makes call processing hard to understand, so I left it out.

You will find that I have put together a document that will allow a person lacking a telecommunications background to understand the basics of GSM. I hope you agree that non-techies are easier to work with after they read this, since they will have a basic understanding of the technical side of GSM.

A Note To Non-Engineers

There is a lot more to GSM than I am introducing in this book. I have tried to provide enough information to give you a basic understanding without giving you a total brain melt down. If you would like more information, talk with the engineers at your local network operator. They are pretty talented folks, and like to explain stuff to those who are interested.

I would also suggest that you not start any arguments with an engineer based upon what you learn here. You may find that I have greatly simplified a process to make it easier to explain. I do think that you will be able to understand a lot more of what the engineers have to say after you read this book.

Building Blocks

GSM Subsystems

Customer Setup & Provisioning

Basic Call Processing

Services and Features

Billing

Future Enhancements

Appendixes

Building Blocks

Stuff you need to know before
the rest of the book makes sense

Chapter 1
Talking The Talk

Years ago when I first got involved with mobile phones, there were not many around. Cellular was just getting started. I found that when I tried to have conversations with the people who worked at the cellular companies, they seemed to be talking a foreign language. Things have gotten better as advanced telecommunications becomes more and more commonplace, but it still can be baffling. In this chapter we will review many of the communications terms that will be used throughout this book.

Mobile telephones are just one of many types of wireless communication; pagers, CB radios and walkie talkies are also wireless. Wireless communications take place using radio signals. Information is sent between the antenna of your mobile phone, for instance, and an antenna on a tower somewhere near you. Information is also sent from the tower back to your phone.

One Way Communication

Communication that allows you to both send and receive information is called "two way" communication. Compare this to listening to your car radio. Information is sent from a radio broadcast tower and is received by your car radio receiver, but your car radio sends nothing back. Communication where information is sent in one direction only is "one way" communication. (By the way, yelling back at the disk jockey does not count as two way communication).

Two way communication, the one where you can talk and listen, is divided into two segments, simplex and duplex. In a simplex conversation participants must take turns speaking. Walkie talkies use a "push to talk" simplex system.

Two Way Communication

In a duplex conversation, like those on current mobile telephone systems, both parties can speak and listen at the same time.

You cannot have a one way or two way radio communication until the information being sent to you is separated from all the other information in  the air. To do this, your transceiver (transmitter/receiver) must be tuned to the correct channel, much like your car radio has to be tuned in before you can listen to your favorite radio station. Luckily, mobile telephone systems find the correct channel automatically.

Electromagnetic transmissions, in this case radio, travel in waves. A wave which has completed one full up and down motion has completed a full cycle. Some waves take longer to complete a cycle than others. The count of how frequently a wave completes a cycle in a second (cycles per second) is actually the frequency of that radio wave. Frequencies are measured in Hertz (one cycle per second) named after Heinrich Hertz, the first guy to ever produce a radio wave artificially.

Since radio waves usually travel in thousands or millions of cycles per second, (Hertz), we use the metric system to make the numbers easier to deal with.

Here's what we end up with:

1 cycle per second = 1 Hertz (abbreviated Hz)

1000 cycles per second = 1 KiloHertz (abbreviated KHz)

1,000,000 cycles per second = 1 MegaHertz (abbreviated MHz)

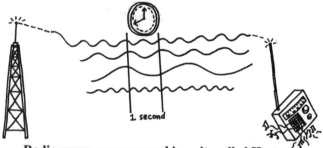

Radio waves are measured in units called Hertz

When you tune your car stereo to 96.9 FM, you are actually telling your radio to look for information being carried on a radio wave traveling at 96.9 MHz, or 96,900,000 cycles per second.

In the United States, the Federal Communications Commission (FCC) allocates radio spectrum in each geographic area.

The definition of geographic areas is quite important, as it is possible to use the same frequency over and over again. Think about watching channel 6 on television in Chicago, and the same station in my home town of Atlanta. The same frequency is used, but the two stations are broadcasting entirely different information.

Before these TV stations could use these frequencies legally, they had to get permission from the FCC. Getting permission usually involves paying a fee. Until recently, Congress maintained that the allocation of radio frequencies must be based upon ability to serve the public. This has changed over the last several years, and it is now legal for the FCC to auction frequencies to the highest bidder. Due to this newfound source of income, the FCC is now flooding the market with radio channels that were previously unavailable to the public. This public auction of frequencies is producing billions of dollars of revenue for the federal government, and new telecommunications systems are popping up everywhere.

The frequencies and their use listed below is by no means complete, but it is a good example of spectrum allocation.

Frequency (Mhz)	Use
26.96 - 27.23	CB Radio
28.00 - 29.70	HAM radio (10 meters)
39.00 - 47.41	Police/Emergency
162.01 - 169.42	U.S. Government
174. 00 - 216.00	TV Channels 7-13
470.00 - 512.00	TV Channels 14-20
824.00 - 829.00	Cellular Transmit
854.00 - 866.00	Private Trunked
869.00 - 894.00	Cellular Receive
1850.00 - 1990.00	Personal Communication Systems

Chapter 2
History

Prior to the introduction of modern day cellular systems, mobile telephones had extremely limited capabilities and low market penetration. The first public mobile telephone system used in the United States, introduced in 1946, was called Mobile Telephone System (MTS). MTS transceivers were bigger than a Baptist church bible and had to be mounted in the trunk of the car. There was no such thing as a handheld unit in this system. In order to place a call you had to first call an operator who then put the call through. The same operator had to monitor your call to determine when the line was free. Talk about a lack of privacy, I bet that operator had the inside story on everything good that was happening in town.

The area in which you can use your mobile telephone is known as the mobile coverage area. All the users of an MTS system shared a single tower in the middle of the coverage area. There were not many radio channels available on the tower, so there were often long waits. Communication on this system was simplex (push to talk), and the coverage and voice quality left a lot to be desired.

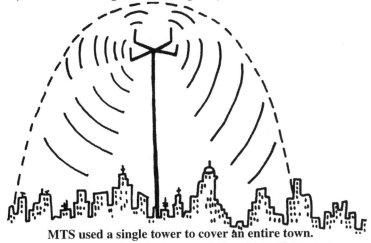

MTS used a single tower to cover an entire town.

In 1965 a new mobile phone system with significant improvements was designed. In keeping with the captivating naming scheme in place, this one was called the Improved Mobile Telephone System (IMTS). While users still shared a single tower, IMTS had many advantages over MTS. Users could direct dial without an operator, so there was a little more privacy, and the systems were now duplexed so you could talk and listen at the same time. IMTS also used a new system known as trunking which automatically allocated the first available radio channel to the next call in line. Trunking made it possible to have more users on the system by assigning the channels for use more efficiently. Despite this improvement, long waits to make a call were not unusual.

The amount of power required to send a signal back to a tower (which could be 40 or 50 miles away) meant that the transceivers still had to be quite large.

Bell Labs, a division of AT&T, started designing a better mobile telephone system. They employed some techniques that had been around for a while and came up with a few new ones on their own. The new phone system was based on four main concepts:

1. The coverage area is divided into smaller areas called cells (thus the name cellular system). Each cell has it's own radio tower, with a radius of approximately 4-6 miles.

2. Radio channels can be used over and over again as long as the same channel is never used in adjacent cells. This process allows significantly more users on the system, even though the total amount of radio spectrum used has not increased.

3. As users travel throughout the coverage area, their conversations are passed from one cell to the next. This process is called a hand off.

4. All of the activities of the network, such as call processing, radio channel assignment and hand-offs are controlled by high speed computer systems.

The first AMPS mobile telephone

Photo compliments of OKI Telecom

The Advanced Mobile Phone System (AMPS) was first trialed in Chicago Illinois in 1978. Since then, the AMPS network has grown to cover the entire country. For spectrum allocation purposes, the U.S. was divided into a series of Metropolitan Service Areas (MSA) and Rural Service Areas (RSA). The FCC issued two licenses for each RSA and MSA. The frequencies used for the AMPS network are between 824 MHz and 890 MHz. At the end of 1996 there were more than 40 million mobile telephone users in the United States. Cellular has become a part of our daily lives.

The AMPS cellular system laid the groundwork for all modern mobile cellular systems. All cellular systems share some common components, although they may have different names.

Switch	connects calls from landline to mobile, from mobile to landline, and from one mobile to another.
Cells	the smaller segments of a divided coverage area.
Cell Sites	radio transceiver tower in each cell.
Site Controllers	computers that control the actions of the cellsites.
Mobile Phones	handheld, transportable, or vehicle mounted.
Databases	storage for information about customers who are allowed access to the network.

While network operators in the United States were building the AMPS networks, the telecom community in Europe was designing and installing their own types of cellular mobile telephone systems. Between 1981 and 1990 there were mobile phone networks built 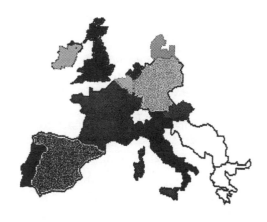 throughout the European continent. The problem was that unlike the United States, the network operators did not all build the same kind of system. Some networks were built based on the Nordic Mobile Telephone standard (NMT); others were based on the Total

Access Communications Standard (TACS) and still others were built on a standard called Radiocom 2000.

This diversity of network types was inconvenient for the customers who used the networks and for the network operators who built them. Users were frustrated because the incompatibility of the networks made it impossible for them to use their phones while they were away from home (roaming). Network operators built smaller systems with a limited customer base which meant it took them longer to see a return on their investment.

The mobile phone operators in Europe believed they should design a single system that met everyone's needs. In 1982 the European telecommunications operators association (CEPT) began working on a new mobile telephone network.

The goals of this new special mobile telephone system that would provide service to the entire European continent (then known as Group Special Mobile) included:

- roaming compatibility - one system for all of Europe

- enhanced privacy - no more eavesdropping

- security against fraud - existing cellular systems consistently lost money to fraud

- enhanced features- why should landline phones get all the new toys

Later, as the project grew in both size and importance, the work was transferred to a group called the European Technical Standards Institute (ETSI). ETSI took on the huge task of documenting the functionality and interaction of every aspect of the GSM network. In order to complete this task a team within ETSI called the Special Mobile Group (SMG) was created.

The best systems design in the world would be of no value if there was no network equipment on which to run it. Network equipment manufacturers were concerned about two issues:

1. How many systems would actually be built?

2. Would all the systems be the same or would there be custom versions for every market?

These concerns became more important as a greater number of countries expressed interested in the Group Special Mobile project (GSM). In 1987 the European Telecommunications Commission oversaw the creation of the GSM MoU or Memorandum of Understanding. A signature on the MoU was an agreement to build a new mobile phone network based exactly upon the specific system design as documented by the Special Mobile Group of ETSI. All nationally licensed mobile telephone operators and regulatory bodies became signatories of the GSM MoU. This guaranteed the equipment manufacturers that there would be a market for the new systems and that all systems would be built according to the same standard. The group of signatories became known as the GSM MoU Association.

GSM quickly became the most talked about new mobile phone system in the world. Many countries outside of Europe began to consider utilizing the GSM standard. Eventually, international demand was so great that the name of the system was changed from Group Special Mobile (GSM) to Global Systems for Mobile Communications (still GSM).

The first test systems were up and running in 1991. In 1992 the first paying customers were signed up for service. It is hard to prove exactly which GSM system operator turned on the very first customer, since systems were

The first commercial network to sell GSM.

brought on-line within hours of each other. It is however, generally accepted that the first paying customer was signed up by Dansk Mobile Telefone in Denmark. The company's product name was Sonofon, which, said correctly, sounds like "So No Phone".

Since that day, more than 200 GSM networks have been deployed in over 100 countries. By the end of 1996 GSM networks world wide were servicing more than 25 million users. The GSM MoU Association is still operational and growing as more network operators sign to become members. The formal objective of the GSM MoU Association is the "promotion and evolution of the GSM systems and the GSM platform."

The original standard, documented by ETSI, is constantly improving and changing. The GSM MoU Association, in conjunction with the Special Mobile Groups of ETSI, approves enhancements and improvements to the GSM platform. The concepts of a published international standard and a constantly evolving common standard are unique to GSM. GSM is unlike most other network architectures, where individual equipment manufacturers create proprietary system deployments.

GSM is a cooperative effort among thousands of the best minds in the world, all of whom share both the workload and the benefits of each others successes.

There are work groups throughout the world specifically designed to allow interested parties to meet and work on finding solutions to systems enhancements that will fit into existing programs of GSM operators. It is interesting to note that, in most cases, these are volunteer groups. MoU member companies pay for their staff to attend work sessions. Some of these groups include:

- Asia Pacific Group (APIG)
- Arab States Interest Group (AIG)
- European Interest Group (EIG)
- GSM North America (GSM-NA)
- Central/Southern African Interest Group (CSAIG)
- East Central Operators Interest Group (ECOIG)
- Personal Communications Network Interest Group (PCNIG)

So what kind of changes are approved by the MoU Association? An interesting example occurred when the Telecommunications Commission in the United Kingdom released spectrum in the 1800 MHz range for use in the new Personal Communications Network systems. The operators who acquired these licenses wanted to build GSM systems, but they had a problem. GSM is specified to operate at 900 MHz and the PCN licenses in the UK were at 1800 MHz. The license holders approached the GSM MoU Association who agreed to allow them to create systems based upon the GSM standard at 1800 MHz. To avoid confusion, they decided to call GSM systems built at 1800 MHz "Digital Communications Systems - 1800" or DCS-1800.

When the US Federal Communications Commission and the Canadian Systems and Industry Group issued spectrum in the 1900 MHz band for Personal Communications Systems (PCS) the MoU Association again agreed to let the GSM standard be built at a different frequency, thus DCS-1900.

Due to increased competition among mobile standards in the US, DCS-1900 is now referred to as North American GSM.

This allows DCS-1900 operators to link themselves more closely to the GSM networks being used around the world.

Name	Frequency	Location Used
GSM	900MHz	Over 100 countries worldwide
DCS-1800	1800 MHz	United Kingdom, Russia, Finland and Germany
DCS-1900 North American GSM	1900 MHz	PCS systems throughout the United States and Canada. [Currently under deployment]

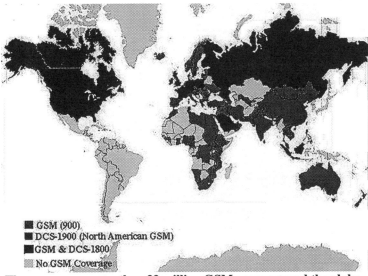

There are now more than 33 million GSM users around the globe.

Chapter 3
Digital vs. Analog Signaling

You hear people extolling the virtues of digital communications everywhere you turn, but what does digital really mean? The mere fact that a system is digital does not make it better than other types of systems. The difference depends on how the system designers take advantage of the capability and flexibility of digital transmission.

The Air Interface

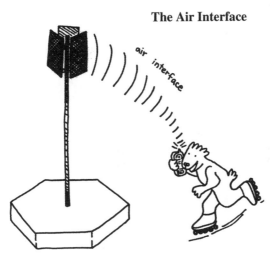

Specifically, when we speak about a digital network we are referring to the manner in which information is sent over the air interface (the air space between the antenna of the phone and the tower). In other words the format in which the information is encoded prior to being sent over the radio transmitter/ receiver device.

AMPS and other mobile phone networks such as NMT-450 or TACS (both used in Europe) use an Analog air interface. The information that is sent over the airwaves is analogous, or exactly equal to, the spoken word. This means that when the signal is received at the other end of the transmission, the original sound is reproduced exactly as it was spoken. Unfortunately any distortion of the original spoken sound such as static, popping or hissing picked up during transmission is also reproduced. Exact reproduction of speech also requires sending a lot of information over the air, this means that a large amount of bandwidth is needed for each call that is in progress on an analog network. The large bandwidth requirement limits the number of users that can be served with a given amount of radio spectrum. Another shortcoming of analog systems is in the signaling capabilities. Computer information commonly required for call privacy or fraud prevention is very difficult to send over an Analog network, whereas a digital network is tailor made for such applications.

GSM, as well as several other technologies, use a digital air interface. More information about these other technologies is included in chapter five. Information is sent over the air as numbers or digits. When you speak, the GSM handset takes small samples of what you say and converts these samples into numbers. This numerical data is compressed together to form a data packet. The packets are then sent over the air to the receiving base station which converts the packets back into a series of voice samples and fills in the blanks. Since human speech follows a very simple pattern, it is easy for a computer to recreate the sound in between the samples and simply play connect the dots. The same process happens in the opposite direction for audio being sent from the base station to your handset.

Speech is sampled by the vo-coder.

The process that GSM uses to sample speech, convert the samples into numerical data and compress the sample is known as encoding. The process of taking an encoded sample and turning it back into speech is known as decoding. The device that encodes and decodes speech is a voice encoder/decoder or vo-coder. There is a vo-coder located in both the GSM network and GSM handsets. With this in mind, it is easier to understand that the higher the number of samples per second taken, which is known as the sampling rate, the more accurate the reconstruction process. In theory, a vo-coder that has a higher sampling rate will produce better sounding audio than a vo-coder with a lower sampling rate. In GSM vo-coders are divided into four types: full rate, half rate and an enhanced version of each. While many believe that the enhanced versions offer improvements in voice quality, this remains a subject of debate in the industry. Half rate vo-coders will allow network operators greater capacity but there are concerns about voice quality. A new type of vo-coder, the adaptive multi rate (AMR) is now being reviewed.

The following chart shows the vo-coder names and their sampling rate:

Enhanced Full Rate (EFR)	26,000 per second
Full Rate (FR)	26,000 per second
Enhanced Half Rate (EHR)	13,000 per second
Half Rate (HR)	13,000 per second

Speech is sampled by the vo-coder and sent over the air as packets of numerical data. These packets are checked for accuracy using a system known as error correction, thus allowing packets that have been corrupted during transmission to be repaired or re-sent. Unlike the analog system, which reproduces any distortion of the original speech, digital encoding allows speech to be transmitted in a format that is crisper, clearer and static free.

Essentially, when compared to analog, digital transmission allows systems to be built that produce the following results:

- Crisper, cleaner, clearer calls - signal distortion is not reproduced.

- Better protection against fraud - more advanced signaling enables us to better validate a subscriber's right to use the service.

- Advanced services and features - advanced signaling allows us to send more information, such as calling line identification, along with the call.

- Better protection against eavesdropping - calls sent over the air are in data packets which can be encrypted for higher security.

- Longer battery life - since a GSM phone compresses speech into small data packets, the transmitter does not have to be turned on as often. The less the transmitter is used, the longer the battery lasts.

Chapter 4
PCS Licenses In North America

Licenses In the United States

In 1993 the US Federal Communications Commission (FCC) allocated a block of radio spectrum to be used for a new type of telecommunications system. According to the FCC, the new spectrum was to be used for:

"a family of mobile or portable radio communications services that involves offerings to individuals and can be integrated with a variety of competing networks."

This new technology was called Personal Communication Services (PCS). The concept behind PCS is that a telephone number is assigned to a person rather than a location. If you want to contact me in today's telecommunications world you need to know where I am and the telephone number for that location. For instance, you need to know that I am in my office and my office telephone number. With PCS you only need to know my personal number, which will reach me whether I am at home, in my car or at the office.

The FCC allocated the spectrum between 1850 and 1990 MHz for new PCS systems. In most cases this band is simply referred to as 1900 MHz. The FCC devised the following scheme to distribute this spectrum. First, the spectrum was divided into 6 blocks. The first three blocks had 30 MHz of spectrum, and the last three got 10 MHz each. The blocks were named A, B, C, D, E and F.

This was the same concept the FCC used to allocate spectrum for AMPS cellular systems in 1984. AMPS was allotted spectrum between the frequencies of 824 MHz - 890 MHz. This spectrum was broken into 2 blocks, A & B. Both original AMPS blocks were issued 20 MHz each. AMPS became so popular that additional spectrum was assigned, bringing the total for each block to 24 MHz.

Unlike the original AMPS project, the FCC did not just give the PCS spectrum away. Rights to use the A-F blocks were auctioned to the highest bidder. Rand McNally (the map people) divided the United States into 48 geographical segments called Major Trading Areas. These segments were created based upon population density and economic factors.

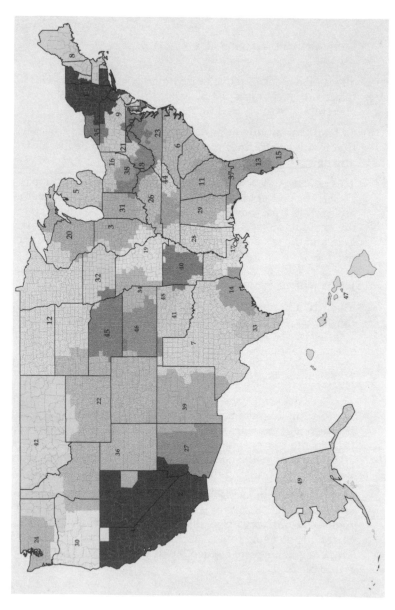

Major Trading Areas

Since it is possible to re-use radio frequencies, the FCC determined that they could sell rights to use both the A and B block spectrum in each one of the MTAs. As an example there is an MTA that includes most of North and South Carolina. AT&T purchased the rights to use the A block frequencies in this MTA and BellSouth Mobility DCS purchased the rights to use the spectrum in the B block. In the neighboring MTA, which includes most of the state of Georgia, AT&T again purchased the rights to the A block, and Powertel, Inc. purchased the rights to the B block. In this manner both the A block and the B block frequencies were sold 51 times each for use across the United States.

The number of potential customers, or POPs, for each MTA was calculated. Interested parties bid in round after round of auctions. The party who bid the highest amount for each POP in a block in each MTA won the rights to offer service using the allocated frequencies in that MTA.

In order to distribute more spectrum and increase competition, the United States was divided into even smaller areas called Basic Trading Areas, or BTAs. There are 493 BTAs, each MTA may have 5 to 10 BTAs within it's borders. The spectrum for blocks C, D, E & F was auctioned in each BTA. The last auctions (blocks D, E & F) concluded in mid January 1997.

Consider the city of Chicago. It is located in an MTA, and a BTA. It is possible that six different companies have purchased the rights to offers PCS service to the same potential customer base. This, by the way, is in addition to the two existing licenses (A & B) already issued in the 800 MHz band for AMPS cellular service. You can imagine the amount of competition there will soon be in the United Sates for wireless service.

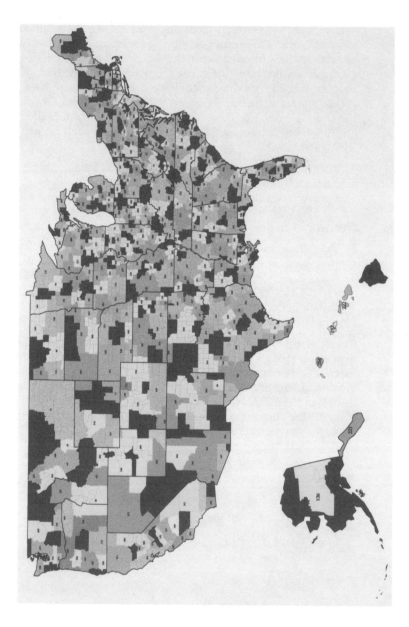

Basic Trading Areas

Two additional rules were created by the FCC to foster competition. The first rule places a restriction on the groups who were allowed to bid on the C and F blocks. This rule encouraged small businesses, businesses owned by women or minorities and businesses owned by native Americans, to participate in the auctions. These designated entities, by definition, leave out the big telecommunications companies.

The second rule addresses the amount of spectrum that any one company is allowed to use in a geographic area.The actual ruling is long and complicated, but simply stated, it means that no one company is allowed to use more than 50 MHz of spectrum in any geographic area. This includes the A-F PCS licenses and the original A & B AMPS licenses.

The chart below shows the breakdown of available spectrum in the U.S.

License Block	Used For	Amount of Spectrum	Sold To
A	MTA	30 MHz	All Players
B	MTA	30 MHz	All Players
C	BTA	30 MHz	Designated Entities
D	BTA	10 MHz	All Players
E	BTA	10 MHz	All Players
F	BTA	10 MHz	Designated Entities

Licenses In Canada

The licensing situation in Canada is similar to that in the United States with just a few differences. The same radio spectrum was allocated for Canadian PCS by their governing body, the Canadian Systems and Industry Group. Frequencies between 1850 MHz and 1990 MHz are divided into the same six groups A-F with the same amount of spectrum allocated to each. The major difference being no division of trading areas. The entire country is one big license area.

Currently there have been 4 nationwide licenses issued.

License Block	Amount	Sold To
A	30 MHz	MICROCELL
B	30 MHz	Reserved For Future Use
C	30 MHz	Clearnet
D	10 MHz	Mobility Canada
E	10 MHz	Reserved For Future Use
F	10 MHz	Rogers Cantel

Chapter 5
Competing Technologies

Unlike the 1984 spectrum lottery, where all systems built used the AMPS Analog standard, the 1993 auctions did not specify which technology would be used to build PCS systems

As a result there are four technologies emerging for use in North America. They are:

- IS-136
- Enhanced Specialized Mobile Radio (ESMR),
- Interim Standard #95 (IS-95/JS-008) and
- North American GSM.

This chapter reviews each of the technologies and compares their strengths and weaknesses.

In order to understand these technologies, we first need to understand the concept of the transport mechanism. Digital mobile telephone systems use different methods to move information back and forth between the antenna at the tower and the antenna of the phone. To illustrate this concept, imagine a large pile of sand that needs to be moved from one place to another. In order to move this sand, we can take one shovel full at a time and carry it to the destination, or we can fill a wheel barrow and move the sand with fewer trips. In both cases we have done the same job, we moved the sand. The method that we used to do the job however, was different. We used a different **transport mechanism**.

Two different transport mechanisms.

With digital systems there are two common transport mechanisms. One is Time Division Multiple Access (TDMA), and the other is Code Division Multiple Access (CDMA).

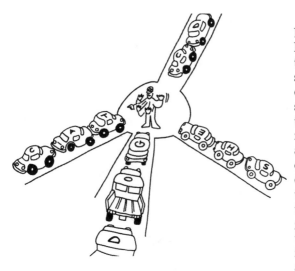

In Time Division Multiple Access, the available spectrum is divided into a series of very tightly defined radio channels, and each channel is divided into time slots. The time slots are grouped together to form frames. TDMA allows multiple users to share the same radio channel by assigning the data packets from each conversation to a particular time slot. As an example: imagine several streets converging into one street. The cars on all of the streets must merge into the single street in order to pass.

In a TDMA network, the base station acts like a traffic cop allowing one car from each street to pass to the single street. When the cop has allowed one car from each street to move forward, he then allows a second car from the first street to pass. In this example, the multiple streets are multiple conversations, cars are data packets from each conversation, the cop is the base station and the single street is the shared radio channel.

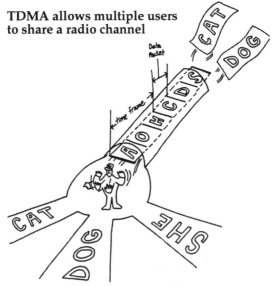

TDMA allows multiple users to share a radio channel

The number of streets converging into one illustrates the number of time slots in a frame. Allowing multiple customers access to the same radio channel by the dividing the channel into time slots gives this transport mechanism it's name, Time Division Multiple Access.

Code Division Multiple Access also works to allow multiple conversations to use the same frequencies, but it works quite differently. CDMA does not divide the available frequencies into smaller radio channels instead, it broadcasts on all available frequencies simultaneously, this method is known as "Spread Spectrum". Each data packet is assigned a special code which allows it to be distinguished from the other packets at the receiving end. As an illustration: imagine that you have just entered a large banquet hall filled with people. At first, all the conversation melds together and what you hear is a roar of sound. As you work your way into the party you find that all the people are actually clumped into smaller groups chatting among themselves. Once you join a conversation you can easily use the particular sound of each person's voice to separate what they are saying from the noise of the rest of the party. Even though everyone is talking in the same room (like broadcasting on the same frequencies) you can use the sound of a person's voice (like a code assigned to that person) to separate what they are saying from the room noise. Allowing multiple users access to the same group of frequencies and separating their conversation by assigning a code to the data packets gives this transport mechanism the name Code Division Multiple Access (CDMA).

Since each person's voice is unique you can pick it out of a crowd

Remember that a transport mechanism like TDMA or CDMA is just the method that is used to move information between the antennas of the base station and handset. It takes an entire network with it's own specific architecture design to get the information that needs to be moved up to the antennas.

IS-136, ESMR and GSM all use TDMA as their transport mechanism. IS-95/JS-008 is currently the only installed mobile phone system using CDMA as it's transport mechanism. Unfortunately, people in the industry tend to refer to system designs according to their transport mechanism. This was fine when there was only one design using each transport mechanism, but as more and more system designs emerge, it tends to make things very confusing. IS-136 for instance is often referred to as TDMA, but GSM also uses TDMA to move information. IS-95/JS-008 is often referred to as CDMA, but what happens when another system using CDMA arrives in the market. If you ask someone what kind of system they are building and they respond TDMA, do they mean that they are building an IS-136 system or some other network that uses TDMA as it's transport mechanism? To avoid this confusion in this book, we will refer to system technologies by their system design names.

IS-136

IS-136 is very similar in network design to the original AMPS networks. The biggest difference is that instead of sending information over the air interface in analog format, this technology uses a TDMA digital transport mechanism. The digital format allows IS-136 to offer several advancements over the original AMPS service, among them cleaner calls with less static, better protection against fraud and eavesdropping and some advanced features such as calling line identification and an internal message delivery system known as short message service. The fact that IS-136 uses TDMA allows it to serve three to five times more customers than an AMPS network could serve using the same number of channels. One of the disadvantages of IS-136, is that the voice encoder-decoder device is not as advanced as those used in other technologies therefore the audio reproduction is not quite as good. A second drawback of IS-136 is that it is based upon an older technology so it's system architecture does not take advantage of all the latest advancements in mobile telephone systems design. Finally, each manufacturer that makes IS-136 systems has their own communications protocol or method of communicating between each part of the system. This means that manufacturer A's base stations will not work with manufacturer B's switch. In other words the network operator has to buy all of it's network components from a single source. Once an operator has chosen it's equipment supplier, there is no competition for the individual parts for the network. Competition for components has two purposes: it drives quality up, and prices down. If you are unhappy with a particular network component, or if your manufacturer's software does not support a particular feature, you are stuck with it.

Since IS-136 is so similar to the original AMPS standard, it is possible to locate radio transceivers for both analog AMPS and digital IS-136 in the same network.

Many AMPS operators are upgrading their network to offer service to new IS-136 customers without shutting off service to their embedded base of analog AMPS handsets. There are also several manufacturers creating handsets that are capable of working on either AMPS or IS-136 networks.

AT&T, who already owns a large group of AMPS systems, is building IS-136 systems in the MTAs and BTAs that they purchased rights to in the PCS auctions. By selling phones

Tower radios can be analog or digital.

that will work on AMPS, IS-136 at 800 MHz and IS-136 at 1900 MHz they will be able to offer service in almost any geographic area of the US with roaming capabilities to any of their other networks. One primary disadvantage of this system is that if a phone locks to an analog AMPS base station, the customer is then susceptible to all the issues of fraud and privacy that led to the creation of digital services in the first place.

ESMR

As its name suggests, Enhanced Specialized Mobile Radio (ESMR), is an upgraded version of existing Special Mobile Radio (SMR) services. SMR services are the communications method of choice for many mobile operations. SMR takes advantage of the automatic channel allocation capabilities of trunking systems to offer access to multiple users with limited radio channels. These systems are most commonly deployed in the 450 to 900 MHz range. ESMR has taken mobile radio based services a step further. ESMR employs a TDMA digital air interface and channel re-use to offer many services previously reserved for cellular systems. These services include fully duplexed two way service, messaging and standard radio dispatch. While not nearly as popular as GSM or IS-136, ESMR has been adopted for deployment in the US and Canada by some very reputable companies. Since there are comparatively few ESMR handsets manfactured, the prohibitive cost of the handset is often the reason more customers don't use this type of service.

IS-95/JS-008

CDMA as an air interface is not new. For many years the US government has been using CDMA for super secret information transmission. What is new is the use of CDMA as a transport mechanism for a public mobile telephone systems. IS-95 is a network architecture that was designed and promoted by a California company called Qualcomm. JS-008 is the specific version of the technology for use at 1900 MHz, much like DCS-1900 is GSM used at 1900 MHz. IS-95/JS-008 is currently the only mobile telephone system which uses CDMA as its transport mechanism. Since IS-95/JS-008 is such a new technology, and has only recently been deployed, not much information is available about actual system performance. Qualcomm, has made numerous promises about the capabilities of the system, and the advantages that a CDMA interface will bring, but many of these claims have yet to be substantiated.

Many people in the industry believe that in the future, CDMA may prove to offer significant advantages over TDMA as the preferred air interface. Many also believe, however, that the IS-95/JS-008 network design will not be able to provide the functionality required to live up to the CDMA air interfaces capabilities. Some functions of the IS-95/JS-008 network such as mobile fax and data have yet to be released to the market. In addition, the systems have not been in operation long enough to substantiate claims of upgraded security, improved speech quality or enhanced features.

The industry is watching closely to see if the IS-95/JS-008 systems can prove that the CDMA transport mechanism is superior enough to TDMA to make it a consideration for adoption by other technologies. Many major telecommunications companies have chosen IS-95/JS-008 as the system with which they will build their PCS networks.

Like IS-136, IS-95/JS-008 uses a proprietary network interface, thus making it impossible to mix and match components from various manufacturers. System enhancements and software updates for an installed system are available to the network operator only from the original equipment vendor at whatever delivery schedule and price that vendor charges. For example, at the time this document was written there was still only one manufacturer of handsets for the JS-008 platform, severely limiting the customers choices. Despite these concerns many major network operators such as Sprint and PCS PrimeCo have implemented IS-95/JS-008 PCS system technology.

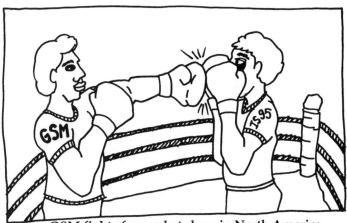

GSM fights for market share in North America.

GSM

As discussed previously, GSM is the most widely deployed of the digital network structures. It is currently in use by more than 40 million users on 200 networks in 100 countries around the world. GSM operates on a proven system platform with more than five years of reliable operation. The open architecture of GSM provides the network operator with the flexibility to buy only the best and most cost effective of each network component required. This also makes it possible to easily integrate the latest advancements in technology. For instance, if the CDMA air interface does prove to be superior, it will be possible for the GSM networks to adopt CDMA as an interface, thus providing the most effective air interface with a proven network architecture.

The advantages of GSM over other types of networks include:

- crisper, cleaner quieter calls
- security against fraud and eavesdropping
- international roaming capability in over 100 countries

- improved battery life (2 to 3 times that of AMPS)

- efficient network design for less expensive system expansion

- efficient use of spectrum (8 to 16 times that of AMPS)

- advanced features such as short messaging and caller ID

- a wide variety of handsets and accessories

- high stability mobile fax and data at up to 9600 baud

- ease of use with over the air activation, and all account information held in a smart card which can be moved from handset to handset

GSM Subsystems

What it takes to make the system work

Chapter 6
GSM Subsystems

Chapter two explained that all GSM networks must be built in accordance with a specifications document created by ETSI. In this book, the parts of the network are divided into a series of subsections. The subsections are:

- The Network Subsystem (NSS)
- The Base Station Subsystem (BSS)
- The Operations Subsystem (OSS)
- The Operations and Maintenance Center (OMC)
- The Mobile Station (MS)

Any parts that do not fit into one of these subsystems are included in a group called:

- Other network components

During the review of the functional parts of each of the subsystems, don't worry about how one part talks to another part, that will come later.

NETWORK SUBSYSTEM

The Network Subsystem includes:

The Mobile Switching Center (MSC)

The Home Location Register (HLR)

The Authentication Center (AUC)

The Visitor Location Register (VLR)

The Equipment Identity Register (EIR)

MSC

The Mobile Switching Center is an advanced electronic telephone switching device. In GSM, like most other cellular systems, the switch is the heart of the network. The MSC has three main jobs:

1. It is a switch. In other words, it connects calls from sender to receiver.

The MSC has three jobs

2. It collects the details of the calls made and received, who called, the time, how long the call lasted, and what features were used.

3. It supervises the operations of the rest of the network components.

For all its power and usefulness, the MSC (and most of the other network equipment) is pretty boring looking. It is a big cabinet of computer cards, power supplies and disk drives. From the outside it looks a lot like the full length lockers you might have had in high school.

HLR

The MSC itself does not contain any records about the phones on the network or the customers who use them. All this information is kept in a series of databases. The first database is the Home Location Register (HLR). The HLR is the central depository for all information required to allow a customer to access and use a GSM network. This database does not store information such as the customer's name or home address, that type of information is stored in the billing system, which will be discussed later.

The Home Location Register

Even if a network operator has more than one MSC, the network usually only has one HLR. There are two situations in which a second or third HLR may be required: the first database reaches maximum capacity or the network operator wants a backup.

Like most other parts of the network, the HLR has a built in backup system. The database is actually divided into two sections, a primary and secondary. If the primary database fails the secondary automatically takes over. Any information written to the primary is automatically written to the secondary. Since they contain exactly the same information at all times, the two sections of the HLR are said to be mirrored. Network equipment that has a built in backup system is called self redundant. In some cases, two of the same devices are installed for redundancy. For additional security, copies of the data in the HLR are stored in a separate location.

In the unlikely case that both sides of the mirrored disk pair are damaged, the stored copies can be used to rebuild the database.

The mobile number is called a MSISDN (Mobile Subscriber ISDN number) and the MSISDN is dialed to call the customer, much like we would use your name to call you across a crowded room.

The GSM network identifies a user by their International Mobile Subscriber Identity Number or IMSI; much like the government identifies a person by their social security number. If someone wanted to call a GSM mobile telephone user, they would dial the customers' MSISDN, if the network wanted to locate that user, it would look for their IMSI.

 The MSISDN can be compared to a users name and the IMSI can be compared to their social security number.

Both the MSISDN and the IMSI are stored in the HLR. In addition to these two numbers there is also a lot of other information that is stored, such as details about which services and features the customer can access. Another number stored in the HLR is the restriction class. The restriction class indicates whether a customer is allowed to place and receive calls and if any restrictions have been placed on the types of calls that can be made. My favorite restriction class is called "Hotline". If a network operator sets an account to Hotline, they can direct outbound calls to a specific number, no matter what number the customer dials. Imagine, you are a little late on your bill, you dial the number to call your very best girlfriend, but instead your call is routed to the mobile telephone company collections department!

Even with the Hotline restriction class active, calls to 911 are usually routed to the emergency services operator.

AUC

Unlike many other mobile telephone networks, GSM is very safe from fraud and eavesdropping. This is because the GSM specifications call for a high degree of security. The Authentication Center (AUC) is the part of the network that is used mainly for security. The AUC is a data storage location, as well as a functional part of the network. The primary data element that is stored in the AUC is called the Ki (pronounced key). There is a unique Ki for every user on the network. No one ever sees this number, it is not written on any paperwork and it is  never transmitted over the air. The AUC uses the Ki to create other security information that is required by the network. The AUC is usually included as a part of the HLR, a sub-database of sorts. So how many AUCs are there? The same amount as there are HLRs. Usually there is only one, but if a network operator has more than one HLR, they will also have more than one AUC.

VLR

Here's an interesting question, if a network operator can have several MSCs but only one HLR, does that mean that the HLR must be consulted for every call? What if there is one switch in Houston and another in Tampa and another in New York and the HLR is in New York? Do all three MSCs have to contact the HLR every time someone uses their phone? The answer is no. There is another database, called the Visitor Location Register (VLR), which handles information about, and approval of, local caller traffic.

The Visitor Location Register

There is a VLR everywhere there is an MSC. The VLR holds records for customers who are using the local MSC that day. In other words the New York VLR only has records about customers who are using the New York MSC that day.

Let's say a customer is using their phone with the MSC in Houston, but the HLR is in New York. To facilitate the process, a temporary record is created for the customer in the Houston VLR. This new record contains much of the same information that is stored in the HLR (MSISDN, IMSI, restriction class and services and features) plus some extra security information. A VLR record is created for every user that uses their service in that market, regardless of whether it is their home market or if they are visiting from another city or another country. You can imagine that if we created a permanent VLR record for every person who used their GSM service in Houston, the database would fill up quickly. To keep this from happening, the information in the VLR is deleted every night.

The VLR has a few other jobs, the first of which is to work with the MSC to ensure that the user really is who they say they are. This validation process is called authentication. GSM networks authenticate carefully and often. Another job of the VLR is to track which customers have their phones turned on, and are prepared to receive a phone call. The last job of the VLR is to periodically update its database on the location of each phone that is turned on and ready to receive calls. This way the MSC does not have to search to find the phone. Keeping up with which phones are turned on, and where they are, is known as mobility management.

EIR

Another database in the network is called the Equipment Identity Register or EIR. The EIR is used to track handsets.

Every GSM handset ever manufactured has a unique serial number. This number is known as an International Mobile Equipment Identity or IMEI. In order to track handsets, we simply put the IMEI of the handset into the Equipment Identity Register (EIR) database. The EIR has three sub-databases:

1. The White List - for all known good IMEIs

2. The Black List - for bad or stolen handsets

3. The Gray List - for handsets that are being tested or evaluated

The EIR is an optional database, it is not essential to have one in order to run your network. A network operator may decide to implement an EIR but only populate one of the database. For

The EIR has three lists; white, grey and black

instance, many operators only use the black list to track lost or stolen phones. If the IMEI of a handset is placed on the black list, that handset will not work on the network. In order to neutralize the possibility for a black market in stolen handsets, the GSM community has implemented an international EIR. Once this database is completely functional a handset stolen in Australia and placed on the black list, could not be used in Australia, or France, or anywhere else in the world. This in effect makes it pointless to steal a GSM handset

TIP: To keep from confusing the IMSI with the IMEI, remember that IMSI has an S for Subscriber. The IMEI has an E for equipment.

The Network Subsystem includes:

- The Mobile Switching Center (MSC) - switches calls, collect s call detail records and supervises the operation of the network.

- The Home Location Register (HLR) - permanent data storage. One HLR can serve many MSCs.

- The Authentication Center (AUC) - Ki storage, security work.

- The Visitor Location Register (VLR) - temporary data storage, authentication, and mobility management There is a VLR at every MSC.

- The Equipment Identity Register (EIR) - holds handset IMEIs in three lists; white, black and gray.

The Network Subsystem

BASE STATION SUBSYSTEM (BSS)

Having all the parts of the network subsystem is useless without a method to communicate over the air with the mobile telephones that are in the market. The parts of the system that are involved with actually communicating over the air are included in the Base Station Subsystem (BSS). The BSS is made up of the Base Station Controller (BSC), the Base Transceiver Station (BTS) and the air interface.

Like all other cellular systems, GSM divides the coverage area into a series of smaller areas called cells. Antennas are strategically placed to serve these coverage areas. Sometimes the antennas are put on towers and sometimes they are bolted to the sides of buildings. There is just no way to guess where you may see a GSM antenna mounted.

BTS

The antenna itself, as well as the radio transceiver that it serves, make up the Base Transceiver Station or BTS. Most GSM networks have several hundred BTSs scattered around the coverage area. The radio transceivers are usually mounted in a box at the base of the tower, or close to wherever the antenna is located.

BSC

The BTS cannot run by itself, it needs a control computer to tell it what to do. That's the job of the Base Station Controller (BSC).   The BSC keeps up with what radio channel is being use by which BTS and what customer is communicating on it. The BSC also regulates the transmit power level of the towers and handsets. As a handset gets closer to the tower, the BSC lowers the transmitter power levels. If a handset travels away from a tower, the BSC raises the transmitter power levels   Unlike other networks, one BSC can support multiple BTSs.   It is not uncommon to see 35 or more BTSs all being controlled by the same BSC.

This is especially handy when it comes time to hand the conversation from one BTS to another BTS in the same area. Chances are both sites are being controlled by the same BSC, and it orchestrates the entire process.

The Base Station Controller

Common Air Interface

The space between the antenna of the BTS and the antenna of the mobile telephone is known as the common air interface. In a GSM network, the communication that takes place over the common air interface is completely digital. Everything that is sent by a GSM network over the common air interface is encrypted with a secret coding scheme that would make Dick Tracy jealous.

OSS

The operations subsystem or OSS contains all the parts of the network that are needed to run day to day operations. The OSS includes the inventory systems, customer care / billing systems and gateways to transport information. The billing system holds all information that is pertinent to a customer's account, such as their name, address, rate plan and what services and features they ordered. Some updates such as posting a bill payment are maintained solely in the billing system. Others such as activating a new account or adding a feature must also be sent to the HLR in order to be effective. When updates are made to a customer's record in the billing system they are automatically passed through a data gateway to the HLR. A data gateway that translates information into the correct format for the HLR is often referred to as a mediation device. It acts as the mediator between the HLR and the billing system. The billing system also contains the rating module that converts call detail records collected by the MSC into information billing records used to create telephone bills that are sent to customers.

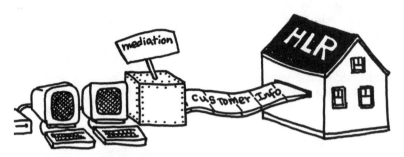

Information is translated by the mediation device

OMC

The operations and maintenance center or OMC is the command center from which every part of the network is monitored. The inside of an OMC looks like a NASA control center with large monitors, computers on every desk and lots of lights and alarms. From inside the OMC network engineers can monitor the function of any part of the network. The system is equipped with alarms to advise the engineers of every sort of failure, from a tower being hit by lightning to a BTS cabinet being vandalized. In most cases the OMC is used to monitor several networks in several cities at the same time. Staffing one central control center 24 hours a day is much less expensive then having engineers on-site at multiple switch locations to monitor operations of each local switch.

Consider this scenario: it is 3:00 AM on Sunday. While most of us are asleep, lightning hits a tower in a cattle field in Houston, Texas. The site is damaged and goes off-line. At the same moment an alarm goes off on a panel in Tampa, Florida where the network operator has their OMC staffed 7 days a week 24 hours a day (7X24). The engineer on duty logs into her terminal and begins to run a series of checks and tests on the site to see what parts are damaged. Ten minutes later she calls the home of the on-call field engineer in Texas, she reports:

> "Site number 36 has been taken down by a lightening hit, I have run the maintenance routines and it appears that the card in slot three of frame two in the BTS is damaged. I have checked the inventory on your truck, and you have the part in stock. The site is completely off-line so this is a priority one call."

In less than half an hour after lightning hit the tower in the middle of the night, an engineer is on his way to the site with advance information of what is wrong and the knowledge that he has the parts to fix the problem. One hour after the tower was hit by lightening, it is back on the air and taking calls. This level of service can be offered for any one of several cities served by that OMC with only one or two engineers awake and on duty in the middle of the night.

The GSM Network

Other Network Components

There are many parts of the network that do not fit into one of the previously mentioned subsections. Examples of other network components include voicemail systems, short message service centers and the executable short message platform. Listed below is a brief description of these components. Each one will be discussed in greater detail later in the book.

Voicemail System (VMS) - The voicemail system allows callers to leave messages, or deliver numeric pages to GSM mobile users.

Short Message Service Center (SMSC) - configures and delivers alpha numeric messages, up to 160 characters in length, to and from GSM mobile users.

Executable Short Message Platform - configures and delivers commands directly to the SIM card over the air (OTA). An example of such a command is the download of a newly activated customer's mobile number. This platform is called the OTA gateway.

Interworking Function (IWF) - processes and formats mobile fax and data. Supports stable, fully error corrected data transmission.

Chapter 7
The Mobile Station

The Mobile Station Subsystem (MS) is made up of two parts, the handset and a smart card known as a Subscriber Identity Module (SIM). The single biggest difference between a GSM handset and any other mobile telephone handset, is that the subscriber information is not programmed into the handset. All the information that is required for a subscriber to gain access to the network is stored in a computer chip on the SIM card. In a radical break from convention, this book will discuss first the SIM card and then the handset.

User information is stored in smart card not the phone

One of the most distinctive characteristics about the GSM system is that customer information is stored on the SIM, not in the handset. This makes it very easy to move your account from handset to handset. It is actually possible to have GSM service and not own a handset. You could just borrow one whenever you wanted to make a call, insert your SIM card, and the borrowed handset is now ready to make and receive calls on your account. Take a moment to consider the implications that this account mobility has for a mobile user: corporations can share from a pool of handsets while maintaining billing records, handsets can be exchanged for repair or upgrade with no impact on the account and travelers have the advantage of carrying their GSM account in their pocket wherever they go.

SIM

The Subscriber Identity Module is a microprocessor chip installed on a plastic card. The memory configuration of the chip is divided into two main areas, one for the operating program, the other for data files used for information storage. The plastic card that the chip is embedded into can be one of two sizes. The first is a full size SIM and is the same size as a credit card. Since the size and chip placement of this card is regulated by the International Standards Organization (ISO) the full size card is referred to as "ISO Standard".

The full size card is simply too large to fit into many of the smaller phones on the market. The second size card, known as a "Plug In", is made by removing most of the plastic from a full size card. The configuration and size of the plug in card is defined in the GSM MoU documentation.

Full Size SIM

Plug In SIM

The SIM card has four primary roles in the GSM system.

1. Authentication
2. Assistance With Voice And Data Encryption
3. Information Storage
4. Subscriber Account Protection

- **Authentication**

The SIM card plays a vital role in proving that a user who is attempting to gain access to the network is truly a valid subscriber, and not a thief using a stolen SIM card. The purpose of the authentication process is to prove that the user has the correct IMSI and Ki. During the authentication process, a random number is sent to the SIM card, this random number and the Ki (which is stored in the SIM) are processed through an encryption algorithm. (An algorithm is a math problem; you put a series of numbers in, crank them around and then you get a new number, called an answer). The answer is sent by the card back over the air to the GSM network. Inside the network the answer is validated, thus proving that the customer has the correct Ki without ever transmitting it over the air. (If we sent this secret information over the air, it could be captured and duplicated by thieves.)

- **Assistance With Voice And Data Encryption**

Each time any information is sent over the air, it must be encrypted in a secure format. To insure that the encryption code cannot be broken, the SIM card runs the Ki and a random number through a second set of algorithms, which results in a different answer, called a Kc. The handset uses the Kc as one of the variables in the encryption process. While the SIM card does not actually do the voice encryption, by providing the Kc, it helps make the process more secure.

- **Information Storage**

The ability to move your account from phone to phone would not be as valuable if all your favorite speed dial numbers, and messages were left behind in your handset. The SIM card is capable of storing your phone list (names and numbers), any messages you may have received, as well as handset preferences such as in what language the prompts will appear.

- **Subscriber Account Protection**

Since the SIM card is, in effect, the customer's account, care must be taken in the event the card is ever lost or stolen.

To prevent a thief from inserting a stolen SIM card in a GSM phone and using the account, users may set a password on the card. The password is known as a PIN (Personal Identification Number), and for additional protection there is a higher level access number known as the Pin Unblocking Key (PUK).

PUK is pronounced "puck" not "puke"!

- PIN

It is possible to disable the PIN function of the card, so that no PIN is required. The PIN number may be changed or deleted by the subscriber. It is as easy as entering your old PIN number and then entering the number you want as your new PIN.

There is a precaution to keep someone from trying different numbers until they figure out your PIN. If the PIN number is entered incorrectly three consecutive times the SIM card becomes blocked. In this case a higher level access number known as a PIN Unblocking Key (PUK) is required.

- PUK

The PUK is used to reset the PIN number. In the event that the SIM has become blocked, you simply key in the PUK code, and then enter a number for the new PIN. This becomes especially useful if the subscriber forgets their PIN number. For additional security if the PUK code is entered incorrectly ten times in a row the card is rendered useless.

Future technology will enhance the applications of the SIM card; not only will your SIM card be an easy way to move your account from phone to phone, but in the near future it will serve other uses as well. You will able to use the same card to charge a meal, make a mobile phone call, rent a video or pay for a plane ticket. Work is in progress to develop multiple application SIM cards, which will provide one or many of these services.

Handsets

The GSM mobile station is a marvel of electronics. In addition to the radio transmitter/receiver device it also contains a voice encoder/decoder device, a display screen capable of displaying alpha numeric messages and many other enhanced features. Since it is a radio transceiver, it must have an antenna. Current handsets have either a retractable or fixed external antenna. Recent trials in Europe are actually introducing handsets with no external antenna at all.

- **Batteries**

The GSM handsets sold in North America are all designed to be used as handheld portable phones, and therefore operate from a battery. The useful life of a battery charge is rated in either standby time, or talktime or a combination of both. Standby time is the amount of time the handset can remain turned on waiting to send or receive a call. Talk time is the amount of time a battery will support a phone call in progress.

There are several types of batteries available for GSM handsets. As discussed earlier, GSM vo-coders sample speech, and transmit these samples in bursts. When the transmitter is turned on it drains the battery. A phone sitting in an idle state uses less battery power than a phone that is transmitting, just as a car that is idling uses less gas than a car moving at full speed. Some types of batteries work better with this burst transmission than others.

The chart below compares different types of battery technology in terms of expense, performance and other notes by assuming that the different batteries are of roughly equivalent physical dimensions.

Type	Expense	Performance	Notes
Alkaline	*Least	Good	Convenient to purchase AA etc. Handsets must be designed to work with alkaline cells
"Nicd" Nickel Cadmium	Below Average	Good	Most commonly used, lowest cost but susceptible to problems with incomplete charging known as the "Memory Effect"
"NiMh" Nickel Metal Hydride	Average	Better	Currently offered in most North American handsets, with no "Memory Effect"

Lilon	**Highest	Best	Smallest, lightest, latest technology. Not widely available due to limited source of supply

* Disposable, so over the long term more expensive
**Prices will drop quickly over the next 24-36 months
***Batteries are actually chosen on the basis of dimension, capacity and voltage. State-of-the-art chemistries, like Lilon, may not yield better performance that extended packs of the more established chemistries, such as NiCd and NiMH.

The signal strength and transmitter power of a GSM handset are not impacted by the battery charge level. A weak battery will not make a call sound bad. If a battery does not have enough charge to maintain a call at the proper power level, the call is discontinued.

It is important to differentiate between handset features and network features. A handset feature is one that is supported solely inside of the handset and is independent of the network. A network feature is a service which is provided by the network or a network peripheral device.

In order for many network features to operate they must be supported by handset functionality. There are also functions of the handset that must have network support to operate. In the following charts this information will be presented as follows:

- **Handset** = Handset Only
- **Network** = Network Only
- **Handset/Network** = Handset Feature with Network Support Required

Network features with no implications to the handset will be discussed in another section.

Feature Name	Residence	Explanation
Antenna Type	Handset	What type of antenna does the handset have, internal, fixed or retractable
Auto Dial # In Short Message Service	Handset/ Network	If network supports Short Message Service (SMS) and a message is delivered with a telephone number in it, the handset will dial the number when the send button is hit.
Auto Redial	Handset	Handset attempts to redial a number when a busy signal is encountered
Battery Charge Time	Handset	Length of time required to charge the battery.
Battery Type	Handset	Type of battery supplied with the phone. See the chart above for more information.
Charger Type	Handset	Options include desktop charger, travel charger cigarette lighter charger etc.
Cigarette Lighter Available?	Handset	Does the phone have a cigarette lighter adapter?
Clock Display/ Alarm Clock	Handset	Time shown on handset display. Does the phone have a built in alarm clock?
Conference Call Supported	Handset/ Network	Ability to link multiple parties on a single phone call.

Feature Name	Residence	Explanation
Dimensions	Handset	Physical dimensions of the handset.
Direct Voice Mail Key	Handset	Single button. When pressed, dials directly into the voice mail system.
Display CLID	Handset	Display telephone number of incoming call.
Display Incoming Name If Stored In Memory	Handset	If incoming calling number matches one stored in memory, either handset or SIM, display the name associated with that number.
Display Size	Handset	Number of lines on display, number of digits that can be displayed per line.
Extended Battery Available?	Handset	Is there a longer life battery available as an option?
Hands Free Kit Available	Handset	Is there a speakerphone car kit available as an option?
Last # Redial	Handset	Redial the last number called by pressing one button.
Maximum Data Rate	Handset/ Network	What is the fastest Baud rate the handset will support for the transmission of fax and data?
Mobile Originate SMS	Handset/ Network	Can the user key a short message into the handset and send it to another user?

Feature Name	Residence	Explanation
Number Of Alert Tones	Handset	How many different alert (low battery, out of service etc.) tones does the phone have?
Number of Ringer Tones	Handset	How many different ringer tones does the phone have?
Other Accessories	Handset	Are there other accessories available with the handset?
PCS Data Card Available	Handset/ Network	Is there a data card available with the handset?
Phone # Storage on handset	Handset	In addition to the storage on the SIM card, is it possible to store names and numbers in the handset, if so, how many?
Receive SMS	Handset/ Network	Is the handset capable of receiving mobile terminated short messages?
SIM Size	Handset	Full size or plug in SIM card.
SMS Storage On Handset.	Handset	In addition to the storage on the SIM card, is it possible to store short messages in the handset, if so, how many?
Speed Dial Quick Keys	Handset	Can handset be programmed to dial a number simply by pressing one key?
Standby Time	Handset	How long will the phone stay turned on and in standby before the battery needs to be recharged?

Feature Name	Residence	Explanation
Talk Time	Handset	How long can a call be in progress before the battery needs to be recharged?
Weight (Ounces)	Handset	Weight of the handset with a standard battery attached.

Other notes about handset features:

Call timers - Many phones are equipped with call timers. There is a temptation to use these timers to calculate how much airtime has been used. There are many factors that contribute to a discrepancy between the amount of minutes shown on an airtime counter and the actual bill.

1. Airtime counter is reset on the same day as the customers bill cycle. In most cases billing periods do not run from the first of the month to the last day of the month. The customer base is divided and billing periods staggered.

2. The airtime counter registers calls in progress. It cannot tell the difference between a recorded announcement, a busy signal and an actual call. The counter will count for many incomplete calls that never appear on the customer's bill.

3. Call waiting or two party call conference use two lines and are therefore charged at twice the rate. The handset, however, only counts one set of minutes.

4. Some network operators offer free call numbers such calls to customer service. The handset will count minutes although the customer is not charged.

CLID On Handsets - In current GSM networks Calling Line Identification only displays the telephone number of the incoming call. The only exception to this is when a call is received from a number that has been programmed into the phone list, in which case many handsets display the name stored with the number. You can have fun with this by loading in the names and numbers of the people who call you most. A phone call from your mother could appear on the display of your phone with the word MOM.

Not all features work on all handsets. Check the features list carefully. Do not assume that all handsets have the features you want.

There are some handset features that were developed and patented by network operators in the United Kingdom. These features were developed outside of the Special Mobile Group, and are therefore not included in the GSM specifications. Such features are grouped into a category known as CPHS which stands for Common PCN Handset Specification. Examples of CPHS functions include Alternate Line Service (ALS) and a special display Icon to show a voicemail message waiting indicator (VMS-MWI).

The next few pages contain pictures and information about GSM handsets. This information was provided to us directly by the handset manufacturers. Not all available handsets are listed, as there are some handsets being sold in North America that we were not provided information about.

Ericsson CA318

Antenna Type	Fixed
Auto Dial# in SMS	Yes
Auto Redial	Yes
Battery Charge Time	Less than 1 hour 20 min
Battery Type	Nickel Metal Hydride
Charger Type	Desktop
Cigarette Lighter Avail.	Yes
Clock Display/Alarm	Yes
Conference Call Supported	No
Dimensions	5.1" x 1.9" x 1.3" (130mm x48mm x 33mm)
Direct Voice Mail Key	No
Display CLID	No
Display Incoming Name if Stored in Memory	No
Display Size	3 line
Extended Battery Avail.	Yes
Hands Free Kit Avail.	Yes
Last # Redial	Yes
Maximum Data Rate	9600 baud
Mobile Originated SMS	No
Number of Alert Tones	5
Number of Ringer Tones	11
Other Accessories	Belt Clip, Batteries, Antennas
PCMCIA Card Avail.	Yes
Phone # Storage on HS	Up to 30 numbers
Receive SMS	Yes
SIM Size	Plug In
SMS Storage on Handset	Yes
Speed Dial Quick Keys	1-9 (Press and hold)
Standby Time	Up to 55 hours
Talk Time	4 hours and 45 minutes
Vocoder Type	EFR & FR
Weight	8.7 ounces

**Assuming normal operating conditions
with battery supplied with phone**

Ericsson CH388

Antenna Type	Fixed
Auto Dial# in SMS	Yes
Auto Redial	Yes
Battery Charge Time	Less than 1 hour
Battery Type	Nickel Metal Hydride
Charger Type	Travel Charger
Cigarette Lighter Avail.	Yes
Clock Display/Alarm	Yes
Conference Call Supported	5 Party
Dimensions	5.1" x 1.9" x 1"
	(130mm x48mm x 25mm)
Direct Voice Mail Key	No
Display CLID	Yes
Display Incoming Name	
if Stored in Memory	Yes
Display Size	3 line
Extended Battery Avail.	Yes
Hands Free Kit Avail.	Yes
Last # Redial	Yes
Maximum Data Rate	9600 baud
Mobile Originated SMS	Yes
Number of Alert Tones	5
Number of Ringer Tones	11
Other Accessories	Belt Clip, Batteries,
	Car Kits
PCMCIA Card Avail.	Yes
Phone # Storage on HS	97 numbers
Receive SMS	Yes
SIM Size	Plug In
SMS Storage on Handset	Yes
Speed Dial Quick Keys	1-9 (Press and hold)
Standby Time	Up to 28 hours
Talk Time	2 hours and 20 minutes
Vocoder Type	EFR & FR
Weight	6 ounces

**Assuming normal operating conditions
with battery supplied with phone**

Ericsson CF388

Antenna Type	Fixed
Auto Dial# in SMS	Yes
Auto Redial	Yes
Battery Charge Time	Less than 1 hour
Battery Type	Nickel Metal Hydride
Charger Type	Travel Charger
Cigarette Lighter Avail.	Yes
Clock Display/Alarm	Yes
Conference Call Supported	Yes
Dimensions	5.1" x 1.9" x 1.1" (130mm x48mm x 28mm)
Direct Voice Mail Key	Yes
Display CLID	Yes
Display Incoming Name if Stored in Memory	Yes
Display Size	3 line
Extended Battery Avail.	Yes
Hands Free Kit Avail.	Yes
Last # Redial	Yes
Maximum Data Rate	9600 baud
Mobile Originated SMS	Yes
Number of Alert Tones	5
Number of Ringer Tones	11
Other Accessories	Belt Clip, Batteries, Antennas
PCMCIA Card Avail.	Yes
Phone # Storage on HS	97 numbers
Receive SMS	Yes
SIM Size	Plug In
SMS Storage on Handset	Yes
Speed Dial Quick Keys	1-9 (Press and hold)
Standby Time	Up to 28 hours
Talk Time	2 hours and 20 minutes
Vocoder Type	EFR & FR
Weight	6.1 ounces

**Assuming normal operating conditions
with battery supplied with phone**

Mitsubishi Wireless G100

Antenna Type	Fixed
Auto Dial# in SMS	Yes
Auto Redial	Yes
Battery Charge Time	1.7 Hours
Battery Type	Nickel Metal Hydride
Charger Type	Desktop
Cigarette Lighter Avail?	Yes
Clock Display/Alarm	No
Conference Call Supported	5 Party
Dimensions	5.31" x 1.89" x 1.22"
Direct Voice Mail Key	No
Display CLID	Yes
Display Incoming Name if Stored in Memory	Yes
Display Size	4 x 12 Characters
Extended Battery Avail.	Yes
Hands Free Kit Avail.	Yes
Last # Redial	Yes
Maximum Data Rate	9600 baud
Mobile Originated SMS	Yes
Number of Alert Tones	
Number of Ringer Tones	8
Other Accessories	Belt Clip, Batteries, HandsFree Kit
PCMCIA Card Avail.	Yes
Phone # Storage on HS	No
Receive SMS	Yes
SIM Size	Plug In
SMS Storage on Handset	No
Speed Dial Quick Keys	Yes
Standby Time	Up to 130 hours
Talk Time	2.2 hours
Vocoder Type	EFR
Weight	5 ounces

**Assuming normal operating conditions
with battery supplied with phone**

Motorola StarTAC Select 8500g

Antenna Type	Retractable
Auto Dial# in SMS	
Auto Redial	Yes
Battery Charge Time	2.5 hours
Battery Type	Lithium Ion
Charger Type	Internal Rapid Charger
Cigarette Lighter Avail.	Yes
Clock Display/Alarm	Yes
Conference Call Supported	No
Dimensions	6.1" cu. in.
Direct Voice Mail Key	2 key quick access
Display CLID	Yes
Display Incoming Name if Stored in Memory	Yes
Display Size	92 x 32 pixel graphical display
Extended Battery Avail.	Yes
Hands Free Kit Avail.	Yes
Last # Redial	Yes, last 10
Maximum Data Rate	9600 baud
Mobile Originated SMS	Yes
Number of Alert Tones	3 call alerts and SMS alert
Number of Ringer Tones	11
Other Accessories	Holster, leather purse pack, handsfree kit
PCMCIA Card Avail.	Yes
Phone # Storage on HS	Yes, 100
Receive SMS	Yes
SIM Size	Full Size
SMS Storage on Handset	No
Speed Dial Quick Keys	Yes, 9
Standby Time	39 - 48 hours
Talk Time	135 - 165 minutes
Vocoder Type	
Weight	3.9 ounces

**Assuming normal operating conditions
with battery supplied with phone**

Motorola Select 6000e

Antenna Type	Retractable
Auto Dial# in SMS	
Auto Redial	Yes
Battery Charge Time	45 min
Battery Type	Slim NIMH
Charger Type	Internal Rapid Charger
Cigarette Lighter Avail.	Yes
Clock Display/Alarm	No
Conference Call Supported	No
Dimensions	10.98 cu. in.
Direct Voice Mail Key	2 key quick access
Display CLID	Yes
Display Incoming Name if Stored in Memory	Yes
Display Size	92 x 32 pixel graphical display
Extended Battery Avail.	Yes
Hands Free Kit Avail.	Yes
Last # Redial	Yes, last 10
Maximum Data Rate	9600 baud
Mobile Originated SMS	Yes
Number of Alert Tones	3 call alerts and SMS alert
Number of Ringer Tones	11
Other Accessories	Various chargers, Data accessories
PCMCIA Card Avail.	Yes
Phone # Storage on HS	Yes, 100
Receive SMS	Yes
SIM Size	Full Size
SMS Storage on Handset	No
Speed Dial Quick Keys	Yes, 9
Standby Time	Up to 45 hours
Talk Time	Up to 200 minutes
Vocoder Type	
Weight	7.4 ounces

**Assuming normal operating conditions
with battery supplied with phone**

Motorola Select 3000e

Antenna Type	Retractable
Auto Dial# in SMS	
Auto Redial	Yes
Battery Charge Time	45 min
Battery Type	AAA NIMH
Charger Type	Internal Rapid Charger
Cigarette Lighter Avail.	Yes
Clock Display/Alarm	No
Conference Call Supported	No
Dimensions	14.34 cu. in.
Direct Voice Mail Key	2 key quick access
Display CLID	Yes
Display Incoming Name	
if Stored in Memory	Yes
Display Size	2 x 12 characters
Extended Battery Avail.	Yes
Hands Free Kit Avail.	Yes
Last # Redial	Yes, last 10
Maximum Data Rate	9600 baud
Mobile Originated SMS	Yes
Number of Alert Tones	2 call alerts and SMS alert
Number of Ringer Tones	11
Other Accessories	Various chargers, Data accessories
PCMCIA Card Avail.	Yes
Phone # Storage on HS	Yes, 100
Receive SMS	Yes
SIM Size	Full Size
SMS Storage on Handset	No
Speed Dial Quick Keys	Yes, 9
Standby Time	Up to 43 hours
Talk Time	Up to 200 minutes
Vocoder Type	
Weight	7.1 ounces

**Assuming normal operating conditions
with battery supplied with phone**

Motorola Select 2000e

Antenna Type	Retractable
Auto Dial# in SMS	
Auto Redial	Yes
Battery Charge Time	45 min
Battery Type	AAA NIMH
Charger Type	Internal Rapid Charger
Cigarette Lighter Avail.	Yes
Clock Display/Alarm	No
Conference Call Supported	No
Dimensions	13.73 cu. in.
Direct Voice Mail Key	2 key quick access
Display CLID	Yes
Display Incoming Name if Stored in Memory	Yes
Display Size	2 x 12 characters, LCD display
Extended Battery Avail.	Yes
Hands Free Kit Avail.	Yes
Last # Redial	Yes, last 10
Maximum Data Rate	9600 baud
Mobile Originated SMS	Yes
Number of Alert Tones	2 call alerts and SMS alert
Number of Ringer Tones	11
Other Accessories	Various chargers, Data accessories
PCMCIA Card Avail.	Yes
Phone # Storage on HS	Yes, 100
Receive SMS	Yes
SIM Size	Full Size
SMS Storage on Handset	No
Speed Dial Quick Keys	Yes, 9
Standby Time	Up to 43 hours
Talk Time	Up to 200 minutes
Vocoder Type	
Weight	6.7 ounces

**Assuming normal operating conditions
with battery supplied with phone**

Nokia 9000 Communicator

Antenna Type	Whip with Joint
Auto Dial# in SMS	
Auto Redial	Yes
Battery Charge Time	2.5 Hours
Battery Type	Lithium Ion
Charger Type	Arbitrative Charger
Cigarette Lighter Avail.	Yes
Clock Display/Alarm	Yes
Conference Call Supported	5 Party
Dimensions	1.5" x 2.5" x 6.8"
	(130mm x48mm x 28mm)
Direct Voice Mail Key	No
Display CLID	Yes
Display Incoming Name	
if Stored in Memory	Yes
Display Size	3 x 10 Characters
Extended Battery Avail.	Yes
Hands Free Kit Avail.	Yes, also built in
Last # Redial	Yes
Maximum Data Rate	9600 baud
Mobile Originated SMS	Yes
Number of Alert Tones	3
Number of Ringer Tones	8; 3 composed by user
Other Accessories	Car Kit
PCMCIA Card Avail.	Not Needed
Phone # Storage on HS	Yes
Receive SMS	Yes
SIM Size	Plug In
SMS Storage on Handset	Yes
Speed Dial Quick Keys	No
Standby Time	Up to 30 hours
Talk Time	120 minutes
	(Talk/Fax/Data Time)
Vocoder Type	
Weight	13.9 ounces

**Assuming normal operating conditions
with battery supplied with phone**

Nokia 2190

Antenna Type	Retractable
Auto Dial# in SMS	
Auto Redial	Yes
Battery Charge Time	1 Hour
Battery Type	NiCD or NiMH
Charger Type	Arbitrative Charger
Cigarette Lighter Avail.	Yes
Clock Display/Alarm	No
Conference Call Supported	Yes (only in handsets manufactured after 3/97)
Dimensions	.9" x 2.2" x 5.8"
Direct Voice Mail Key	Yes
Display CLID	Yes
Display Incoming Name	Yes
if Stored in Memory	Yes
Display Size	5 x
Extended Battery Avail.	Yes
Hands Free Kit Avail.	Yes
Last # Redial	Yes
Maximum Data Rate	9600 baud
Mobile Originated SMS	Yes
Number of Alert Tones	4
Number of Ringer Tones	8
Other Accessoriess	Colored covers, vibrating battery
PCMCIA Card Avail.	Yes
Phone # Storage on HS	125 memory locations
Receive SMS	Yes
SIM Size	Plug In
SMS Storage on Handset	No
Speed Dial Quick Keys	Yes
Standby Time	Up to 22 hours
Talk Time100 minutes	
Vocoder Type	
Weight	8.3 ounces

**Assuming normal operating conditions
with battery supplied with phone**

Nortel 1911

Antenna Type	Fixed
Auto Dial# in SMS	Yes
Auto Redial	Yes
Battery Charge Time	1.5 Hours
Battery Type550	Nickel Metal Hyrdide
Charger TypeTravel Charger	
Cigarette Lighter Avail.	Yes
Clock Display/Alarm	Clock Only/No Alarm
Conference Call Supported	
Dimensions	5.9" x 2.3"
Direct Voice Mail Key	No
Display CLID	Yes
Display Incoming Name	
if Stored in Memory	No
Display Size	
Extended Battery Avail.	Yes
Hands Free Kit Avail.	Yes
Last # Redial	Yes
Maximum Data Rate	9600 baud
Mobile Originated SMS	Yes
Number of Alert Tones	3
Number of Ringer Tones	4
Other Accessories	Car Kit, Desktop Charger Leather Case
PCMCIA Card Avail.	Soon
Phone # Storage on HS	No
Receive SMS	Yes
SIM Size	Full Size
SMS Storage on Handset	No
Speed Dial Quick Keys	Yes
Standby Time	Up to 24 hours
Talk Time	1hour 40 minutes
Vocoder Type	FR
Weight	7.9 ounces (225 grams)

**Assuming normal operating conditions
with battery supplied with phone**

Sagem CS 610

Antenna Type	Retractable
Auto Dial# in SMS	Yes
Auto Redial	Yes
Battery Charge Time	90 Minutes
Battery Type	Nickel Metal Hydride
Charger Type	Travel Charger or Desktop
Cigarette Lighter Avail.	Yes
Clock Display/Alarm	Yes
Conference Call Supported	6 Party
Dimensions	6.1" x 2.3" x.07" (130mm x48mm x 28mm)
Direct Voice Mail Key	Yes (PRG Key) Programmable
Display CLID	Yes
Display Incoming Name if Stored in Memory	Yes
Display Size	4 x 12 Characters
Extended Battery Avail.	Yes, High Capacity NiMH
Hands Free Kit Avail.	Yes
Last # Redial	Yes, Last 20 numbers
Maximum Data Rate	9600 baud
Mobile Originated SMS	No
Number of Alert Tones	15
Number of Ringer Tones	18
Other Accessories	Belt holster, Hands Free, Antennas, Serial Data Cable, Soft Case
PCMCIA Card Avail.	Yes
Phone # Storage on HS	
Receive SMS	Yes
SIM Size	Full Size
SMS Storage on Handset	No
Speed Dial Quick Keys	Yes, Press Number and#
Standby Time	Up to 40 hours
Talk Time	120 minutes
Vocoder Type	
Weight	7.7 ounces

**Assuming normal operating conditions
with battery supplied with phone**

Sagem CS 635

Antenna Type	Retractable
Auto Dial# in SMS	Yes
Auto Redial	Yes
Battery Charge Time	90 Minutes
Battery Type	Nickel Metal Hydride
Charger Type	Travel Charger or Desktop
Cigarette Lighter Avail.	Yes
Clock Display/Alarm	Yes
Conference Call Supported	6 Party
Dimensions	6.1" x 2.3" x.07" (130mm x48mm x 28mm)
Direct Voice Mail Key	Yes (PRG Key) Programmable
Display CLID	Yes
Display Incoming Name if Stored in Memory	Yes
Display Size	4 x 12 Characters
Extended Battery Avail.	Yes, High Capacity NiMH
Hands Free Kit Avail.	Yes
Last # Redial	Yes
Maximum Data Rate	9600 baud
Mobile Originated SMS	Yes
Number of Alert Tones	15
Number of Ringer Tones	15, + Vibrating Ringer
Other Accessories	Belt holster, Hands Free, Antennas, Serial Data Cable, Soft Case
PCMCIA Card Avail.	Integrated Data Transmission in Handset
Phone # Storage on HS	100
Receive SMS	Yes
SIM Size	Full Size
SMS Storage on Handset	Yes
Speed Dial Quick Keys	Yes
Standby Time	Up to 40 hours
Talk Time	120 minutes
Vocoder Type	
Weight	7.7 ounces

**Assuming normal operating conditions
with battery supplied with phone**

Siemens 1050

Antenna Type	Retractable
Auto Dial# in SMS	Yes
Auto Redial	Yes
Battery Charge Time	2 hours
Battery Type	Lithium Ion
Charger Type	Rapid, constant current
Cigarette Lighter Avail.	Yes
Clock Display/Alarm	No
Conference Call Supported	Yes
Dimensions	6.2" x 2.36" x1.14"
Direct Voice Mail Key	Yes
Display CLID	Yes
Display Incoming Name if Stored in Memory	Yes
Display Size	1.35" x 1.91" fully graphical
Extended Battery Avail.	Yes
Hands Free Kit Avail.	Yes
Last # Redial	Yes
Maximum Data Rate	9600 baud
Mobile Originated SMS	Yes
Number of Alert Tones	
Number of Ringer Tones	5
Other Accessories	Extended battery, Desktop charger
PCMCIA Card Avail.	Yes
Phone # Storage on HS	No
Receive SMS	Yes
SIM Size	Full Size
SMS Storage on Handset	No
Speed Dial Quick Keys	Yes, 49
Standby Time	Up to 27 hours
Talk Time	Up to105 minutes
Vocoder Type	EFR
Weight	7.0 ounces

**Assuming normal operating conditions
with battery supplied with phone**

Siemens 1250

Antenna Type	Retractable
Auto Dial# in SMS	Yes
Auto Redial	Yes
Battery Charge Time	2 hours (80%), 4 hours (95%)
Battery Type	Lithium Ion
Charger Type	Standard transformer type charger
Cigarette Lighter Avail.	Yes
Clock Display/Alarm	No
Conference Call Supported	Yes, 5 callers plus user
Dimensions	5.6" x 1.9" x1.0"
Direct Voice Mail Key	Yes, icon based
Display CLID	Yes
Display Incoming Name if Stored in Memory	Yes
Display Size	5 lines x 16 characters, full graphic, COLOR display
Extended Battery Avail.	No
Hands Free Kit Avail.	Yes
Last # Redial	Yes
Maximum Data Rate	9600 baud
Mobile Originated SMS	Yes
Number of Alert Tones	12 indicator icons, 17 sound alerts
Number of Ringer Tones	10 tones, 5 levels of loudness
Other Accessories	Car kit, Belt clip, TTY
PCMCIA Card Avail.	Yes
Phone # Storage on HS	No
Receive SMS	Yes
SIM Size	Plug-In
SMS Storage on Handset	No
Softkeys available	2 double sided keys
Speed Dial Quick Keys	Yes, 49
Standby Time	85 hours (minimum)
Talk Time	4.25 hours (minimum)
Vocoder Type	EFR
Voice Memo	Yes, up to 20 sec. record time
Weight	5.6 ounces

**Assuming normal operating conditions
with battery supplied with phone**

Handset Personalization

The value of cellular handsets in the North American market has been seriously degraded by a practice known as handset subsidy. For many years, AMPS cellular agents and dealers were paid sizable commissions and residuals for signing up new cellular customers. To entice customers to sign up for service many dealers used a portion of their commission to offset or buy down the cost of the handset. In most AMPS activations in North America today, a $200.00 - $300.00 handset is given away in conjunction with a long term service contract.

One of the unique features of GSM is the ability to activate customers and make any required changes to their SIM cards without having to manually program the handset. Since there is no programming, and no electronic serial numbers to read, GSM handsets are easily sold at mass market department stores. This eliminates the requirement for agents and dealers, and enables an "off the shelf" purchase. The elimination of the need for manual programming opens up opportunities for new distribution channels. Easily accessible locations to purchase handsets are one of the keys to success in this competitive marketplace.

To compete with AMPS phones that are practically given away, however, GSM network operators must buy down the cost of the handsets. At the time this book was written, GSM handsets in North America were being purchased at wholesale cost of $400-$550 and being sold at retail outlets for prices ranging from $150-$200. As you can see, network operators have a significant investment in each handset.

Network operators know that they will see a return on their investment in the handset as long as the customer who purchases it signs up for service and uses the phone on a regular basis. The problem is that GSM, by nature of the SIM card, allows users to swap cards and handsets at their convenience. This means that a customer could purchase a subsidized phone from GSM carrier "A" and insert a SIM card from GSM carrier "B" and it should work. GSM carrier "B" gets the advantage of an active customer with little or no acquisition costs while GSM carrier "A" has effectively lost their investment in the handset.

To avoid potential losses of this kind, many North American GSM operators have chosen to personalize the handsets that they sell. In other words the handsets are programmed to accept only SIM cards issued by the operator who sold the handset. With this program, multiple users can share a handset purchased from carrier "A", but a SIM card from carrier "B" would be rejected by carrier "A"s handset.

This personalization can be disabled by a code which is unique to each handset. Network operators may negotiate with each other to unlock phones, these are typically handled on a case by case basis.

Health and Safety Issues - The growth of mobile telephony has raised questions about health and safety issues. The growing concern over having a tower in someone's "back yard" has made local zoning committees hesitant to grant approvals for new tower installations. Other concerns include interference with pacemakers, hearing aids and hospital equipment. Work is in progress to create directional antennas that will reflect the RF energy away from the head and body.

Towers - Concerns about the safety of towers can be addressed by looking at the facts. GSM radio towers are built to the strictest safety standards for both radio emissions and general safety. Most towers are designed to collapse on themselves rather than fall over, so there is no worry about them falling on a house. Recent studies also prove that the proximity of a cellular tower had little or no impact on the resale value of a home.

The chart below illustrates data that should alleviate fears about radio emissions:

Household Item	Field Density*
Nintendo game transformer	1,510 mG
Computer monitor (desktop)	489 mG
Hair dryer (at one inch)	110 mG
Face of a digital clock	17 mG
PCS Tower at 100 Feet From Base	.017 mG

*mG is a unit of measure for magnetic field density. (The G stands for Gauss, the man who invented the unit of measure). **Source: Zapata Engineering, P.A. Charlotte NC**

Hearing Aids - Because of the pulsed nature of digital handset transmissions, there may be interference with particular hearing aids. This same type of interference can also be caused by fluorescent light bulbs, computer systems and electric motors. Although this type of interference is not specific to GSM, the GSM community has been especially active in research to find ways to alleviate problems caused by the interference. The problem is definitely one of proximity, the phone must be within approximately one foot of the hearing aid to cause interference. Devices such as headsets or special plug in cables are currently being marketed to allow those with hearing impairments to enjoy the advantages of advanced mobile communications.

Pacemakers - It has been proven that digital devices have the potential to cause interference with pacemakers. The digital transmitter must be in very close proximity to the pacemaker (1-4 inches), which does not commonly occur in day to day life. Individuals who have pacemakers should exercise simple precautions when using a GSM phone or any other digital device. They should not hold a GSM phone directly to their chest. Even this should not cause a threat to life. Pacemakers step in to regulate the heartbeat only when required. Though the signal from a digital device might confuse the pacemaker, it should not interfere with normal heart functions.

Hospital Equipment - Most hospitals limit the use of cellular phones, paging systems and two way radios because there is a possibility for interference between hospital equipment and these types of devices. Most manufacturers are in voluntary compliance with RF shielding guidelines for the manufacture of hospital equipment. In order to avoid problems, always consult the hospital equipment manufacturer before using any type of two way radio device in any proximity to hospital equipment.

(Airlines also restrict the use of mobile telephones due to concerns of interference with navigational equipment.)

Customer Setup & Provisioning

Now That I Have A Network, What Do I Do With It?

Chapter 8
Ordering SIM Cards and Setting Up the HLR

As you have guessed by now, the SIM card is a vital part of the GSM system operation. While the internal file structure of the SIM is defined by GSM, what is placed in those files varies greatly among network operators. The information included in your SIM has a huge impact on the way that the SIM card works in conjunction with both the network and the handset. For instance, a network operator in San Diego may wish to use English and Spanish as their handset display language choices, while a Canadian operator may use French and English. What the handset display shows when the phone is powered on, how many speed dial numbers a user is able to store and what number the handset dials when the user presses the voice mail button are some of the issues that must be considered when developing the SIM card file layout, or the electrical profile.

In addition to the information stored in the card, outside design on the card and whether the serial number should appear on the front or the back of the card are other elements of the SIM profile. The appearance of the card includes both the artwork, and the graphics (bar codes etc.) used to identify the card. These issues must be taken into consideration when creating the SIM cards' graphical profile.

Careful investigation in choosing a SIM card supplier, and a correct set up of card profiles will pay dividends in both the short term and the long term. SIM cards cost the network operator approximately $9.00 to $13.00 each to create. (This price is for an 8k card, 16k cards may cost significantly more.) This is a relatively low dollar item, but it is not the place to cut corners to save money. Any money saved cutting a corner on card profile creation or manufacturing will quickly be lost many times over in money spent on returns and customers service.

Once a card manufacturer is selected and the profile is completed the next step is to place a card order. The order is placed by creating an electronic file which is usually placed on a standard 3 1/4 inch floppy diskette. The order disk, as it is known, is created by the network operator and contains one IMSI and one serial number for every card to be manufactured. The disk also specifies which profiles are to be used in manufacturing the cards, and whether the cards should be full size or plug in.

The order disk is sent from the network operator to the card manufacturer, who manufactures the card bodies, inserts the chip into the card and then prints the primary artwork on the card. After the cards have been manufactured, they are personalized. The serial number is printed on the outside of the card and the chip is loaded with the IMSI and any other information as specified by the electrical profile. It is at this stage of personalization that the Ki number is created. An individual Ki is generated by the card manufacturer to be used in conjunction with each IMSI supplied by the network operator. The IMSI-KI combination is written to the SIM card and to the diskette. The information on the diskette is encrypted and can only be decoded by the card manufacturer and the network operator both of whom have a secret numerical key.

To put this a different way, the Ki is created, loaded into the SIM, loaded onto the diskette and encrypted by a computer. During the manufacturing process, no human being ever sees the IMSI-Ki combination. This is true for the process of loading the information into the network as well.

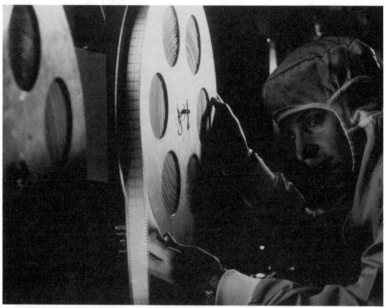

A technician prepares microchips to be inserted into SIM cards
Photo compliments of Gemplus Card Int'l

Before the SIM cards can be used on the network, information on the cards must be loaded into the Home Location Register, or HLR.

The information loaded into the HLR to activate a subscriber includes:

- The IMSI - the number that uniquely identifies the user to the network. The IMSI is stored in the SIM card as well as the HLR.

- The MSISDN - better known as the mobile number. The MSISDN identifies the user to the landline telephone network. It is the MSISDN that you dial to reach a GSM mobile customer.

- Services and Features - which network based features such as call waiting or short messaging the customer is allowed to use.

- Restriction Class - restrictions on the types of calls a customer can make. Common restrictions include:

  No restrictions - ready to make and receive phone calls.

  Suspended - service has been temporarily disabled.

  Hotline - no matter what number is dialed by the user (except 911), route all calls to a designated number. Such as collections for past due accounts.

Additional information is placed into the Authentication Center (AUC), a sub-database of the HLR. The IMSI-Ki combination is stored in the AUC. The Ki information on the SIM manufacturer diskette is decoded as it is loaded into the AUC, and then immediately re-encoded using a different encryption scheme.

In many cases the network operator may elect to "pre-activate" a SIM card prior to issuing it to the customer, or in fact before they even know which customer will receive the card. Pre-activations is the key to "off the shelf" distribution. In this case information is loaded as follows:

- IMSI - from manufacturer diskette.

- Temporary MSISDN - since it is not known what area this card will be sold in, we cannot insure that the proper area code and prefix will be selected. To avoid problems a temporary place holder is loaded (a fake MSISDN), until the real number is issued.

- Services and Features - can either be left blank or pre-loaded with a standard feature set which can be modified later.

- Restriction Class - in most cases the restriction class is set to Hotline, with all calls routed to a new service activations representative.

There is virtually no security risk in this process. No calls can be placed because a fake MSISDN was used and it would not be recognized by any wireless or landline telephone system. Any outbound call attempt would simply be routed to the new customer activations group.

Chapter 9
Distribution

Now that the SIM cards have been pre-activated they are ready to be shipped for sale to prospective customers. Each network operator decides on a method to ship SIM cards, handsets and accessories. Some operators have their own distribution warehouses. Others choose to turn this task over to a distribution specialist, thus freeing the operators' resources to concentrate on the task of managing a successful mobile telephone network.

There are several equipment distributors in North America. Their services range from simple box moving to kitting, packing, shipping, inventory management, and even warranty returns processing.

Another decision for the network operator is whether to ship the handsets and SIM cards separately, or to package them together as a GSM service kit. This kitting process can be handled by the network operator, or contracted out to a distribution vendor. In North America, where most handsets are subsidized by the network operator, the SIM card is often shipped already inserted into the handset.

Once the cards and phones or both are packaged with additional promotional information provided by the network operator, the phones are ready to be shipped for sale to the end user. The method of sale varies from company to company but, in general, the sales team can be broken into a series of channels:

Direct Sales - network operator employees who sell handsets and services directly to the public.

Indirect Sales - a series of authorized sales agents (also known as dealers). Indirect sales channels are usually paid on a commission or residual basis.

Operator Owned Retail Stores - retail stores or kiosks which are owned and staffed by network operator employees.

Indirect Retail Sales - retail stores not owned by the network operator. GSM handsets and SIM cards can now be purchased in retail facilities ranging from department stores to health clubs.

In most cases indirect retail stores buy handsets wholesale and sell them at retail prices. There is not usually a commission involved with this channel.

Telemarketing - telephone sales groups who either make outbound calls to solicit sales or answer incoming calls resulting from advertising or promotional campaigns. Customers who purchase handsets or service via this channel receive their products directly from facilities known as customer fulfillment centers. Like other forms of distribution, some network operators do their own fulfillment, while others hire distribution specialists for this job.

Sales channels may carry handsets, handsets kitted with SIM cards, or mobile telephone accessories. In some cases sales channels sell a SIM only, thus allowing a customer who already owns a GSM handset to sign up for service. (Remember that some handsets are locked to their original system operator. See Handset Personalization in Chapter 7.)

Chapter 10
Activation

Once the customer receives a handset and a SIM card the next step in the process is to activate their account. GSM does not require that the handset be programmed therefore activation can take place over the telephone. For convenience sake, some activations are completed for customers while they wait. This on-site activation is common at a network operator owned retail location, where there is an activation terminal in the store. Whether the activation takes place in person or over the phone, the steps in the process are more or less the same.

Customer Details

General customer information is entered into the Customer Care and Billing System. This includes the customer's name, social security number and billing address. This information along with optional passwords for later service changes are entered into the billing system and an account is created for the customer.

Credit Check

A credit check is run on the applicant in order extend credit to the potential customer for service. In the United States, there are several agencies which specialize in maintaining records of, and creating reports about, an individual's credit history. Many network operators have automatic links to a credit agency with scoring systems built into the billing systems. The scoring system looks for certain indicators, both positive and negative, in a potential customer's credit record. Some systems use these indicators to create a credit score for the potential customer. Other systems simply return a pass or fail indication.

Many network operators offer customers who do not pass a credit check three choices:

1. Pay a deposit - the customer is offered the option to pay a service deposit. The amount of the deposit depends on the score returned by the credit agency. Many automatic credit scoring systems will return a suggested amount when a deposit is required.

2. Prepay - Many GSM networks and/or billing systems allow the subscriber to prepay for services prior to using them. A customer who makes an advance payment of $200.00, for instance, is allowed to use the network until they accrue a bill equal to $200.00, at which point their service is interrupted until

they make another payment. Prepay systems can be supported either in the GSM network or in the billing system. Work is underway to develop special disposable prepaid SIM cards, that will work like a prepaid long distance phonecard.

3. Return the handset to where it was purchased - this is not an attractive option for anyone involved, the retailer loses their margin and must handle a return, the network operator loses a potential customer and the customer does not get mobile service. Equipment returns are expensive and take a lot of manpower to process. Early estimates are that as many as 15% of handsets sold through indirect retail channels are returned. Efforts to develop programs to reduce this number will pay large dividends.

Assigning The MSISDN

One of the most unique features about the North American GSM process is that a MSISDN (mobile number) can be assigned over the phone at the time of activation. Many other cellular systems have tried to implement pre-activation programs. Since all other cellular systems require the number to be physically programmed into the handset, the biggest stumbling block has always been getting the pre-programed phones to the appropriate geographic areas. You can imagine the inventory headaches: not enough phones with Dallas numbers and too many with Madison numbers, Milwaukee got three phones with Florida numbers in them, and Florida needs phones with a specific Boca Raton exchange.

With GSM the relationship between the customer and the mobile telephone number is built in the HLR, not in the handset. During the activation process, we only need to determine from which area the customer wants a number and then assign it to them. A relationship between the IMSI in the customer's SIM card and the new mobile number assigned to them is then created in the HLR.

A Special Note On IMSI Ranges - Many North American GSM system operators have won licenses for more than one MTA or BTA or a combination of both. Winners of these licenses are awarded a special number, which is known as a Mobile Network Code (MNC). In North America this is a three digit number, in other parts of the world the MNC has only two digits. The number plays a very important role in helping other networks determine a customer's home network when they are roaming. The MNC is a part of the IMSI: IMSI = MCC + MNC + MSIN (IMSI = Mobile Country Code + Mobile Network Code + Mobile Subscriber ID Number).

Network operators who have more than one market must decide whether to use a single MNC for all markets or a unique MNC for each market served. This decision has a huge impact on marketing, operations and sales. From a marketing and operations standpoint it is much easier to have a single number, but this may have adverse effects in network planning.

Rate Plans

As a part of the activation process the customer must choose a rate plan, and any extra services and features. A rate plan is simply the combination of monthly service fees, price per minute for airtime and any services or airtime that may be packaged with the plan. Choice of rate plans and features may be restricted depending on the results of the credit check. For example a customer with poor credit may not be authorized to direct dial long distance calls.

The methods used to explain the rate plans and services and features vary by company, but usually tend to be a combination of printed collateral material, information the customer has from prior experience, or advice from a salesperson or customer service representative. Rate plans and services pricing are discussed in greater detail in Chapter 18.

The rate plan is not loaded into the HLR. Rate plan information is used by the billing system to determine how much to charge for services used. It is necessary to activate any features that are included as core services in the HLR. Many companies choose to activate all core features and services when the account is set up during pre-activation.

Features

As discussed earlier in the handset section, North American GSM subscribers can choose from a vast array of services and features. Some of these features are based solely in the handset, some are based in the network and some must be supported in both places. The services and features available for GSM subscribers are discussed later.

Information about the features chosen must be added to both the billing system, so that the customer can be properly billed, and the HLR, to allow the services to be used on the network.

Changing The Restriction Class

In most cases, pre-activated accounts are set up with the restriction class set to Hotline. This routes all calls made from the pre-activated phone directly to the activations center. After the customer account information has been established, the restriction class is usually reset to "Normal", which makes the phone ready to send and receive calls.

Downloading Information To The HLR

To facilitate the process, and ensure synchronicity between the billing system and the HLR, all the aforementioned changes are made in the billing system, and then downloaded to the HLR. Since most billing systems do not format the information in a manner that is understandable by the HLR, the information must pass through a translator. As previously mentioned, this translations device is called a gateway, or a mediation device. Once the information from the billing system has been translated by the mediation device, it is sent to the HLR which updates the account. The customer account is now fully active and ready to send and receive phone calls.

Over the air update - SIM

If the process ended here, the customer account would have been fully activated over the phone, which in itself is a giant step forward in convenience. To make the customers' service even more convenient many network operators have chosen to use the ability of GSM to modify the contents of the SIM card over the air. While there are many files on the SIM card which can be updated over the air, there are only three files that are commonly updated during activation (not all operators modify all three and not all networks can support this function.)

- The MSISDN - it is not necessary for the MSISDN to be in the SIM or the handset for the phone to work on a GSM network. As a matter of fact, the first version of

GSM SIMs did not even have a place to store the MSISDN. (This was quickly changed when customers began to ask for the ability to recall their mobile telephone number from their handset.) As a part of the activation process, most North American system operators send the newly assigned MSISDN over the air to be stored on the SIM card.

- Greeting Message - while not as common, some GSM network operators change the message that is displayed when your handset turns on. As an example, if I signed up for service and registered my account in the name of George Lamb, the message "Hello, George Lamb" could show up on the display of my phone every time I turned it on.

- Services Table - to make the handset menus as easy to use as possible GSM supports a special table on the SIM card called the services table. The handset only displays menu instructions for those features which are listed as active in this table. In other words, if a customer did not subscribe to Call Forwarding, the handset would not display menu instructions on how to make call forwarding work. Most operators set up their SIM services table with a standard set of features when the cards are activated, and then simply update this table if the customer decides to pay for extra features. (It is possible to bypass this functionality by marking all features as active when the SIM is manufactured.)

Basic Call Processing

Ok, Ok, So How Does It All Work

Chapter 11
Selection & Authentication

Once the activation process is complete the customer will of course want to use their telephone. This section of the book describes the basic processes which must take place to allow the handset to make and receive phone calls. A detailed technical description of this process could fill volumes. This section, more than any other in this book, has been greatly simplified and is intended to give the reader a basic understanding of the calling process.

Network Selection & Registration

As soon as a GSM handset is turned on, the handset begins searching a pre-specified group of radio channels looking for a GSM network. Since it is possible that the handset may find more than one network in the area where the phone has been powered up, the handset uses a set of tables in the SIM card to help determine which network to request service from, and which ones not to bother with. These tables are the PLMN tables. PLMN stands for Public Land Mobile Network. Each mobile telephone network has a PLMN number assigned to it. The PLMN number is made up of two numbers, the Mobile Country Code (MCC), and the Mobile Network Code (MNC). You might recognize these numbers as being the first numbers of the IMSI.

Public Land Mobile Network = Mobile Country Code + Mobile Network Code (PLMN = MCC + MNC)

The Preferred PLMN table includes the networks that the mobile should request service from before trying others. The Forbidden PLMN table includes the networks in which a handset has been previously denied access. In order to keep these tables updated, the PLMN value of a network that denies service is automatically added to the Forbidden PLMN table.

When the handset has found a network from which to request service, it "Registers" with that network by sending a message to the network. The message is a request for permission to use the service and it includes the customer's IMSI.

Registration

VLR Check

The MSC itself does not contain any tables with information about who is or is not allowed to use the network. It must request this information from the Visitor Location Register (VLR). If the VLR has information about the IMSI which has requested service then the process of authentication begins. If there is no information about the IMSI in the VLR, the network must request approval from the HLR prior to granting service to the user.

Checking the Visitor Location Register

HLR Validation

The serving network must then contact the Home Location Register (HLR) which contains the customers account information. Communication between GSM network components such as MSCs and HLRs all over the world is completed via high speed data messages in a format known as Signaling System Number 7 (SS-7).

This thought leads to a question. If a customer can turn on their phone and use their service on any GSM system in the world, how does the serving switch know which HLR to request information from? If a customer from Texas uses their service in Scotland, how does the MSC in Scotland know where to find the users account information? The answer is in the IMSI format:

$$IMSI = MCC + MNC + MSIN$$

In our example, the switch in Scotland would look at the Mobile Country Code to determine that the subscriber was from North America, and then at the Mobile Network Code (MNC) to determine from which operator's HLR to ask permission.

Then the serving network sends a request over the data network and asks for permission to offer service to the user.

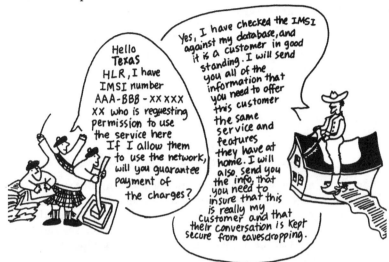

Getting approval from the Home Location Register

After receiving approval from the HLR, the serving system sets up a record in the local VLR for that user. That record remains in the VLR until approximately midnight when the database is cleared out. As long as the record is in the database, the serving system has all the information required to offer service to the customer.

The VLR record setup includes:

1. The MSISDN - identifies which mobile number is associated with the IMSI.

2. The IMSI - used to identify the mobile user.

3. Services and features - a list of services and features that the customer has paid for.

4. Restriction class - status of the account.

5. The "Triplets" - at the time of HLR approval, the AUC provides three numbers used for security. Since the numbers are always provided together, they are referred to the "Triplets". The numbers are:

   - Rand - a random number
   - SRES - the signed response
   - Kc - a mathematical derivative of the Ki

Authentication

Once the VLR record is set up, the serving system must verify that the user requesting service is in fact a valid user and not a "bad guy" attempting to make fraudulent calls with that number. The process used for this validation is called Authentication. Authentication takes place each time the phone is turned on, and on every incoming or outgoing call.

Verification that the Ki stored in the AUC of the HLR matches the one that is stored on the SIM card completes authentication. One important issue is that the authentication process take place without sending any secret information over the air. The ability to trap secret information sent over the air, you may recall, was what led to fraud in the original AMPS cellular network.

There is no easy way to explain Authentication, but put simply, the AUC takes the Ki and a number that it chooses at random and runs them through a mathematical program called the A-3 algorithm. The answer that results is called the signed response. The random number (Rand) and the Signed Response (SRES) are then passed to the serving VLR.

When requested by the network, the SIM card takes the Rand (provided the VLR) and the Ki (stored in the card) and runs these two numbers through the same A-3 algorithm. The resulting answer (SRES) from the SIM card is sent to the VLR and compared to the SRES created by the AUC. If these two numbers match, we have proven, without ever sending the original secret Ki over the air, that the Ki stored in the AUC matches the Ki stored in the SIM.

**GSM validates the Ki without
ever sending it over the air**

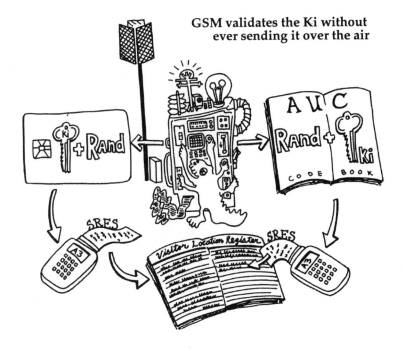

Chapter 12
Call Processing & Encryption

To understand how the system processes phone calls, we must review several scenarios. These scenarios include outbound calls from the handset (mobile originated), calls made to the handset (mobile terminated) calls made from landline phones to mobile phones (land to mobile), calls made from mobile phones to landline phones (mobile to land) and calls made from one mobile phone to another mobile phone (mobile to mobile). First I will explain the steps taken to route calls and then examine the encryption process used to ensure that calls are kept free from eavesdropping.

Mobile To Land

Calls made from a mobile telephone to a landline are the easiest to explain. The mobile telephone transmits a call request message, which includes what number is being called. The MSC validates that the mobile telephone is on the approved list in

A mobile to land call

the VLR, and then completes authentication. Now the MSC is ready to route the call. First it checks the number to see if it should be routed to another mobile telephone user on the same network. Since the number dialed was not a mobile number, the GSM network passes information about the call to the Public Switched Telephone Network (PSTN). The PSTN validates the number, verifies that the call can be delivered and connects the mobile telephone user with the landline number dialed.

Land to Mobile

Call processing of a landline phone calling a mobile telephone is not as easy. The PSTN receives the call request and looks at the number dialed to determine where to send it. Since the number dialed is among those assigned to a mobile telephone user, the landline telephone call sends information to the home MSC for the user being called. The home MSC queries the VLR to see if the VLR has a valid record for the dialed telephone number.

A valid VLR record would indicate that the user has the telephone turned on in their home market.

A land to mobile call

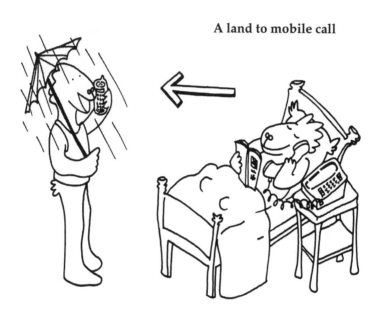

Home market

Once we have verified that the user is in their home market, call routing can continue. The MSC advises the VLR that there is an incoming call for a particular MSISDN. The VLR responds with the IMSI associated with that MSISDN, as well as the last known location of the mobile user. The system validates that the telephone is turned on and ready to receive a call, and authenticates the receiving SIM to prevent fraud. Finally, the network instructs the handset to ring and display the incoming Calling Line ID of the calling party.

Callers don't have to know where you are to reach you

If the customer accepts the call, a connection is established between the landline party and the mobile party.

"Foreign" market

If the user is not in their home market, the home network attempts to locate the customer so that the call can be routed. It does this by querying the HLR which contains information for the MSISDN being called.

The home MSC would send a query to the serving MSC in Houston Texas. If the switch in Houston advised that the mobile was turned on and ready to receive a call, the call would be routed to Texas. After authentication, in Texas, the mobile customer's handset would ring and if the call was accepted, a connection would be made between the landline caller and the mobile customer.

 It doesn't matter whether you are using your service in Paris, Texas or Paris, France because all GSM networks in the world are designed to talk to each other, your call can be routed directly to you.

A mobile to mobile call

Mobile To Mobile Call

A call routed from one GSM phone to another GSM phone on the same network never leaves the network. The calling mobile telephone transmits a call request message which includes the number of the mobile phone being called. The MSC validates that the calling mobile telephone is on the approved list in the VLR, and then completes the authentication step to insure that this is not a fraud attempt.

The MSC queries the VLR to see if it has a valid record for the mobile phone being called. A valid VLR record indicates that the user being called has the telephone turned on in that market.

The MSC advises the VLR that there is an incoming call for a particular MSISDN. The VLR responds with the IMSI associated with that MSISDN, as well as the last known location of the mobile user. The system validates that the telephone is turned on and ready to receive a call, and authenticates the receiving SIM to prevent fraud. Finally, the network instructs the called handset to ring and display the incoming Calling Line ID of the calling party.

If the customer accepts the call, a connection is established between the two mobile telephones.

Unanswered Calls

As with any other type of mobile telephone network, there are several options for calls that are not completed:

1. Allow the calling party to hear ringing until they get tired of waiting and hang up. This is common with landline phones, although not very efficient as it ties up ports in the mobile switching center.

2. Route to a recording that advises that the call could not be completed. This is common in AMPS cellular systems, but it does not allow the caller to leave a message.

3. Route to a voice mail system or live answering bureau. This is the alternative used by most PCS network operators.

Encryption using A5/A8

Security was very high on the list of priorities for the team that designed the GSM network. Security includes not only protection against fraud but also against eavesdropping. This issue is especially important for current analog cellular systems. While it is highly illegal, it is also quite easy to modify commercially available radio scanners to monitor the conversations on analog networks. The process of scrambling voice and data sent over the air is known as encryption.

 The advanced encryption process makes it virtually impossible to eavesdrop on calls on a GSM network.

Before tackling the encryption process, let's review what we learned in Chapter 3. GSM phones do not transmit all of what you say over the air, instead your voice is sampled, and these samples are compressed into data packets. GSM uses Time Division Multiple Access (TMDA) therefore the data packets from your call are mixed up with data packets of several other calls. This sampling and mixing process makes it impossible to listen to a GSM conversation with a standard scanner. The encryption process insures that it will be impossible to eavesdrop on a GSM network no matter what equipment is used.

Every data packet that is sent over the air in the GSM network goes through the following encryption process:

1. The SIM card generates a special number known as the Kc by running the Ki and a random number (Rand) through the A8 algorithm.

2. The Kc is passed from the SIM card to the handset (this step allows a derivative of the Ki to be used without disclosing the actual Ki).

3. The data packet, the Kc and the current number of the TDMA frame are all run through the A5 algorithm which produces a numerical answer.

4. This answer is slotted into the TDMA frame and sent over the air to be used for authentication.

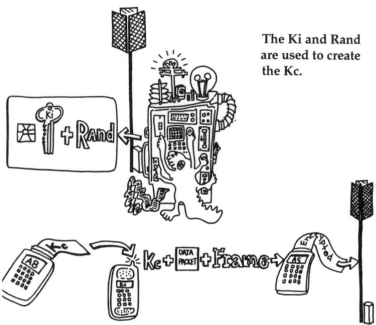

The Ki and Rand are used to create the Kc.

Call data is sent through a double blind encryption algorithm

As if this were not enough, GSM also uses a process called frequency hopping which means that the first data packet may go out on radio channel 1, and the next on channel 5 and the next on channel 3. To make the coding and encryption process easier to remember, I often use this example:

There is a chain of diners throughout the South and one of their specialties is hash brown potatoes. They serve these potatoes many different ways, you can get them **scattered, smothered** and **covered**. To remember all of the encryption stuff just think of it this way:

- We chop the calls into little pieces just like hash browns

- We **scatter** the parts of the call across multiple radio channels

- We **smother** them with encryption algorithms

- We **cover** our butts by never sending secret information over the airwaves

Chapter 13
Mobility Management

The GSM system takes advantage of a high speed data network to "look ahead" and determine whether or not a call or message can be delivered. Sending queries over the data channel is very cost effective. Forwarding a call over a voice channel ties up expensive transmission facilities. The system must keep up with which mobile phones are turned on and active in the network area at all times. To make call delivery as efficient as possible the system also tracks the last known location of a mobile phone. Towers are divided into groups by geographical location. These groups are called location areas. When a call comes in, the system looks for the handset in the last known location area. This limits unnecessary communications with other towers in the network, thus making the call delivery quicker and more efficient.

The process of keeping up with which phones are turned on, and which location area they are in is commonly referred to as "Mobility Management". Mobility Management uses three primary processes: Attach, Detach and Periodic updates.

Attach

Each time the handset is turned on, it finds a network to use (Network Selection) and requests permission to use the network (Registration and Authentication). The mobile network now knows that the phone is turned on and on which tower it is operating.

Detach

As a part of the power-down process the handset advises the network that it is shutting off. In other words each time you turn the handset off, it pro-actively detaches itself from the network.

Periodic Updates

There are situations in which the handset could be powered down without advising the network (detaching). The battery going dead or being removed, or the handset being dropped in the water or damaged are just two such situations. To insure that the handset status information is current, the network contacts the handset from time to time to insure that it is still powered up. If the handset responds the network simply updates the last known location. If the handset does not respond, the network assumes that the handset is shut down and lists it as detached. Periodic updates also insure that the location area of the handset is kept current.

Services and Features

Yeah, But What Can I Do With It

Chapter 14
Basic Features

North American GSM offers a myriad of services and features including all the basic features that are currently offered with analog cellular services, as well as some advanced features made possible by the use of the digital air interface. While all the features and services described in the following chapters, are available with GSM, not all network operators have chosen to implement all the services and features discussed. If a feature or service can be used, it should work essentially the same everywhere; but not every service or feature is available in all markets.

Call Waiting

Call waiting is a network based feature which must also be supported by the handset. With this feature activated, a user who is on a call will receive an audible beep to alert them that there is a second incoming call. Incoming calls can be accepted, sent directly to another location such as voicemail or simply rejected. (The caller on a rejected call hears a busy signal). Once the waiting call is answered, most network operators charge airtime for both calls.

Unlike AMPS cellular and most landline phones, most GSM handsets will display the incoming calling line ID of the waiting call. This allows to user to see who is calling prior to accepting the call.

Call Hold

Call hold is a network based feature that must also be supported by the handset. The call hold feature allows a user to 'park' a call. When the call is parked, the calling party hears silence, and the user can take other calls or hold conversations that cannot be heard by the caller.

Users can select call barring features

Call Barring

Call barring is a network based feature which is activated either by the network operator, or in some cases by the user via the handset.

(Call barring initiated by the network operator for credit reasons, such as toll restriction, cannot be disabled from the handset.)

Call barring has many different versions which are used for different reasons (Not all GSM networks support all these features):

All Calls

- Bar All Calls - bars all calls to and from the mobile except for emergency calls. Note that short messages can still be sent and received with call barring activated.

Outgoing Calls

- Bar All Outgoing Calls - no outgoing calls (except emergency) from any market area.

- Bar International Calls - no outgoing international calls from any market area.

- Bar International Calls Except To Home Country - no international calls except to home country.

- Toll Restriction - Local calls only. This is a network based restriction. It can only be changed by the network operator and is not technically considered a part of the GSM call barring set.

Incoming Calls

- Bar All Incoming Calls - no incoming calls.

- Bar Incoming When Roaming - no incoming calls when using service outside of the home market. This service is used to prevent toll charges associated with forwarding calls from the home market.

Call Forwarding

Call forwarding is a network based service that can be activated from the handset. Call forwarding allows calls to be sent to various numbers under conditions defined by the user. When they are turned on, most GSM handsets will remind the user that call forwarding services are activated. Many network operators charge airtime for calls forwarded to other numbers.

Call Forwarding Unconditional - Activated by the user or the network operator. The home switch does not attempt to route any calls to the mobile phone, but instead routes all calls directly to the designated number. This is activated by dialing *21* + the "forward to" number + #. The service is deaticvated by dialing #21#.

Call Forward Conditional - Activated by the user or network operator. It allows calls to be forwarded only under certain criteria,

No Answer, Busy or Unreachable. Each criteria can forward calls to a different number: for instance busy calls to go to the secretary while unanswered calls could go to voicemail. In most cases conditional forward calls are sent to a voice mail service.

- No Answer - phone rings but is not answered.
- Busy - phone is busy (call waiting is off, or there is already a call holding).
- Unreachable - phone is turned off or outside the service area.

It is possible to forward calls to the same number under each of these conditions without having to set them individually. To do this dial **004* + the "forward to" number + #.

Multi Party Call Conferencing

Multi Party Call Conferencing is a network based feature that must be supported by the handset. This service allows users to initiate multiple calls and link all of the calls together. Most network operators charge airtime for each call initiated by the mobile user.

Chapter 15
Advanced Features

Calling Line ID

Whenever possible, the GSM handset displays the originating telephone number (the caller's number) on incoming calls. In order for the caller's number to be delivered to a GSM handset it must be received from the originating network. In some cases, such as older phone networks which do not support Calling Line ID delivery or calls from analog mobile phone services, this is not possible. It is also possible for the calling party to block the delivery of their number; in which case the display on the GSM handset reads "Private."

The GSM network also delivers the caller's mobile telephone number (MSISDN) with each outbound call. It is possible for the GSM user to block the delivery of their telephone number, either on a per call basis or permanently.

Per Call Block - users can decide to block the delivery of their telephone number on a per call basis. To block delivery, users dial *67 and the number to be called. For instance *67 713-2919 would call 713-2919 but not deliver the number of the mobile calling party.

Per Line Block - users can request that their number not be delivered with any outbound call. In the event that per line block is activated, users can force their number to be delivered by dialing *82 prior to the number. For instance a caller with per line block activated could dial *82 713-2919 when the called party received the call, the MSISDN of the caller would be displayed.

 Many landline telephone users who have caller ID are used to having both the name and the telephone number of the calling party delivered. While we are very close to being able to offer it, this feature is not yet supported by North American GSM, which currently delivers the telephone number only. Most handsets support matching incoming Calling Line ID to numbers stored in memory. This means that if my home telephone number along with the name 'George' is stored in your handset, each time I called you from home, your phone would match the Calling Line ID to the stored number and 'George' would appear on your handset display.

Alternate Line Service

It is possible to have more than one phone number associated with a single SIM card. This is called alternate line service (ALS). A user may decide to use one number for personal calls and the other for business calls. If all outbound business calls were made on one number and all outbound personal calls were made on the other number, accounting is greatly simplified for the user. ALS is a network based feature that must also be supported in the handset.

A user with ALS can switch between numbers to make outbound calls, but cannot use both lines at once. For instance, there is a caller on line 1 and an incoming call on the same number, the user will hear a call waiting beep. However, if a caller is on line 1 and a call comes in on line 2, this will not activate call waiting. When a call is not in progress, an incoming call on either line will cause the phone to ring. Most handsets are capable of supplying a distinctive ring tone depending on which number the call is coming in on. A user can tell by the ring tone which line the call is coming in on.

About Two Phones One Number: There are many questions about two GSM phones sharing the same MSISDN at the same time. The answer is GSM does not support two phones sharing the same phone number. The same enhanced security that hampers fraud prevents multiple users on the same number. There are some network providers who can support this functionality using special network devices known as Advanced Intelligent Network (AIN) platforms, but these services are not very common.

AMPS mobile phone customers often get two phone numbers on the same phone by having their original phone cloned. In other words the mobile number and the ESN (electronic serial number) from the customer's phone are reproduced on a second AMPS handset. This process is exactly the same as that used by thieves who wish to use a valid subscribers service fraudulently. The FCC clearly states that the duplication of an ESN is illegal:

"It is a violation of section 22.919 of the Commissions rules for an individual or company to alter or copy the ESN of a cellular telephone so that the telephone emulates the ESN of any other cellular telephone. Moreover it is a violation of the Commissions rules to operate a cellular phone that contains an altered or copied ESN."

In June of 1996, after AirTouch Cellular filed a lawsuit, the Federal Court in Santa Ana California issued an order for the Cellular Extension Company to shut down operations. Not only did the company have to shut down, but they were also required to turn over the records of their customer base. How would you feel if the federal authorities dropped by to talk to you about your extension cellular phone?

Interestingly enough, it is not just a legal issue with GSM. A GSM network will not allow two SIMs to operate with the same account numbers. If the network detects the use of a duplicate

Tracking cloned phones

card anywhere in the world it will shut down both the original and the duplicate. This is a comforting thought, especially if your analog cellular phone has ever been cloned.

Closed User Group

Closed User Group (CUG) is a network based service which allows members of the group to call each other by simply dialing the last four digits of the MSISDN. Many network operators offer significantly lower billing rates for calls between members of a CUG. Groups must be established by the network operator. In most cases, each member of the group is charged a small monthly fee for the benefit of being in a CUG.

Advice Of Charge

Advice Of Charge (AOC) service is designed to present users with a tally of the actual costs of phone calls. In theory, users are able to see the cost of the last call, or the total spent since the last time the counter was reset. This is especially desirable in roaming situations where it is difficult to know exactly how much the foreign carrier charges. While AOC is often a topic of conversation in the GSM community, it is rarely, if ever, made available for customer use on the network. The service has had many stumbling blocks in it's implementation. Problems with exchange rates, security and matching rates with billing systems have delayed the availability of this service.

Fax & Data

A virtual office by the sea

"Virtual Office" and "Mobile Professional" are new terms that are being used more and more today. GSM allows the mobile professional to sit in a virtual office and make calls, send faxes, read e-mails or even "surf the net." A virtual office can be at home, in the car or on the beach. By connecting a laptop computer to a mobile handset, GSM users have access to almost all of the same features and services that are available via landline telephone services. This includes data transmission, fax, e-mail, and internet services.

A special adapter card is needed to use a computer with a GSM phone. This adapter card is called a PCS Data Card and is designed to fit into any laptop computer or portable fax machine that can accept a standard type II PCMCIA card. (PCMCIA stands for Personal Computer Memory Card Industry Association.) A PCMCIA card is a credit card sized computer modem designed to fit into a laptop computer or portable fax machine.

A PCS Data Card

To use mobile fax or data, a user plugs the PCS data card into the computer or fax machine and connects it to a jack in the handset via a connector cable supplied by the card manufacturer. GSM supports all the latest updates (Group III) in fax functionality, as well as most data formats.

The only restriction with GSM and mobile data is that the speed at which information can be sent to or from the computer or fax machine is limited to 9600 bits per second (Baud). (Faster baud rates are currently being standardized.) Some systems, with special compression techniques, can operate faster, but this rate depends on which handset is used. While this is not as fast as many landline modems, GSM mobile data services are considerably faster, and more stable than those offered on previous mobile telephone networks, and wireless packet data networks.

 TIP: Think of the data transmission rate (BAUD Rate) of a computer or fax machine like a water hose. The smaller the hose the longer it takes to move a gallon of water. The lower the baud rate the longer it takes to transmit a fax, receive an e-mail or transmit a file. GSM transmits information at up to 9,600 pieces (bits) of information per second. It takes a lot of bits to send a fax. An average page of fax is broken down into approximately 600,000 bits. At a rate of 9,600 bits per second, the page would take just over one minute to transmit.

Mobile fax and data must be supported in both the handset and the network. The actual modems that are used to format the information are located in the network. Like voice, all fax and data information sent over a GSM network is secured against eavesdropping or interception by a complex encryption process.

Many network operators charge additional fees for the use of mobile fax and data, and it is usually necessary to have additional telephone numbers assigned for incoming faxes or data. The most common telephone number assignment scheme is listed below:

First MSISDN .... Inbound and outbound voice,

Outbound Fax and Data

Second MSISDN....... Alternate Line Service

Third MSISDN........ Inbound Data

Fourth MSISDN....... Inbound Fax

This multiple telephone number plan may seem awkward, but it is necessary to insure that incoming calls are processed properly.

Voice, fax and data must be sent over the air in a particular format:

Voice - voice is sampled by the vo-coder prior to being encrypted and sent over the air. Voice does not need to go through a special modem.

Fax - all of the data provided is used to reconstruct a fax, it therefore cannot be sampled by the vo-coder, but must be sent in its entirety. Fax information is structured in a special "Transparent Error Correction" format by a network modem.

Data - like fax, data too must bypass the voice encoder and be sent in it's entirety. Unless the receiving party is an ISDN number, data information must be formatted in a special "Non-Transparent Error Correction" format by a different type of network modem.

On outbound calls the handset is able to tell the network what type of call is being processed. In this manner the network knows how to process the call: fax and data bypass the vo-coder and are sent to the appropriate modem device, voice is sent directly to the vo-coder to be sampled. This is why it is possible to route all three types out over the same MSISDN.

On inbound calls, the GSM network has no way of knowing the format of an incoming call. Therefore the MSISDN is assigned in such a way that the network knows what to expect from each line. For instance all incoming calls on the first and second MSISDN are voice, all incoming calls on third are data and all incoming calls on the fourth are fax. The network would then know to route all calls on the first and second number to the vo-coder but to route all calls on the third line to the data modem. All of the calls on the fourth line would be routed to the fax modem The actual assignment of number groups for each call type is left up to the network operator.

Roaming

Roaming allows a GSM subscriber to use service in many markets other than their home market. These other markets may be operated by the same network operator or different network operators. Since all GSM networks are designed and operated based upon the same standard, the services and features that a customer uses at home can "follow" them from market to market, indeed from country to country.

Charges incurred while roaming are presented to the customer on their regular home carrier bill. Records containing data about calls made, or received, are collected in the market where the services are used. If a customer uses their phone on a network operated by a carrier other than their home carrier, the serving carrier calculates the amount due for these services and presents the data records along with a bill to the customer's home carrier. The home carrier pays the serving carrier and in turn bills the customer for the services used.

For a network operator to offer service to a customer from another market the operator must be assured that they will be reimbursed by the user's home carrier. This assurance is pre-negotiated in the form of a reciprocal contract, or a roamer agreement. The primary points of a roamer agreement include:

- rate at which airtime will be charged
- rate at which long distance will be charged
- method by which billing data will be exchanged (tape, disk, etc.)
- format in which billing data will be presented
- method by which payments will be made

To facilitate the process of exchanging roamer records, special data handling companies known as clearinghouses have been established. The clearing house serves as both a data record validater and a banking institution. It is the job of the clearinghouse to insure that all records are in a format that can be easily read and re-billed by the home carrier. In addition, the clearinghouse acts as a financial moderator between the two carriers.

For example, a customer from Washington travels to Texas and uses $100.00 worth of airtime. A customer from Texas travels to Washington and uses $105.00 worth of airtime. All records are formatted and billed by the clearinghouse. The net result of which is that the Texas operator owes the Washington operator $5.00. Payment of the $5.00 "Net Settlement" is significantly easier than processing checks for the total amount due each other.

The role of the clearinghouse becomes even more important when a customer from the United States, for instance, travels to France. The data record format that is used in North America is different than that used in other parts of the world. (The North American format is known as TAP II, the rest of the world uses TAP I.) The clearinghouse must not only translate the data records into the appropriate TAP format, but also handle the issue of currency exchange and any applicable taxes or tolls.

The ability to allow national or international roaming is dependent on both the establishment of roaming agreements, and the ability of the carrier to process roamer records. Most North American GSM carriers are now offering North American roaming, and widespread international roaming capability is expected by the end of 1997.

Chapter 16
Voicemail

Most mobile network operators now offer voicemail services. A voicemail platform is nothing more than a combination phone switch and computer. Each user in the system is assigned a virtual mailbox for which they record greetings. Messages are routed to the appropriate mailbox based upon the number that the calls are forwarded to, or caller ID of the number the call was forwarded from. When a call is routed to the voice mail system, the system retrieves the greeting and then plays it to the caller. In most cases the greeting prompts the caller to leave a message. This message is sampled and digitized, much like a GSM phone does with voice, and the data is written to storage on the computer. When the mailbox owner calls to check for messages, the system retrieves the digitized message, reconstructs it and plays it back.

If a message is left on a voice mail system from a GSM phone, the sound of the original spoken word has already been sampled, digitized and reconstructed once by the GSM system. The voicemail system then samples the message, digitizes it again and stores the sample. If the owner of the voicemail service calls on a GSM phone to check messages the whole sampling and digitization process happens yet again. Each time the original voice message is sampled and digitized, it loses a little bit of it's original character. For this reason, leaving voice mail for yourself is not a good way to test the voice quality of digital mobile phone systems.

Voicemail boxes services are offered with different capabilities. Some of the factors that can be modified on a voicemail box include:

Length Of Greeting	Length of the message that the caller hears.
Length Of Message	How long a message the caller can leave.
Number Of Messages	How many messages can be stored in the mailbox before it is full.
Days Of Storage - New	How long the system will store a new (unheard) message before it is automatically deleted.

Days Of Storage - Old   How long the system will store a previously heard message before it is automatically deleted.

Most GSM network operators in North America are taking advantage of the advanced capabilities of modern voicemail systems, in conjunction with the enhancements brought by GSM, to offer additional services to the user. GSM Conditional Call Forwarding is being used to automatically route calls to voicemail anytime the telephone is turned off, busy or the caller rejects the call.

Once inside the system, callers are offered the option of leaving a voice mail, entering a return telephone number or in some cases, being transferred to a live operator to leave a message. In each case, the mailbox owner is alerted that a message has been left by using GSMs ability to send alpha numeric messages directly to the handset. This alpha text capability is called Short Message Service (SMS). See chapter 17 for more information on SMS.

In addition to using the Short Message Service to alert the user that messages are waiting or to deliver numeric pages, it is also possible to send Message Waiting Indicators (MWI) directly to a traditional paging system. Customers can receive a page each time a message is left for them.

Chapter 17
Short Messaging

At first glance GSM Short Message Service (SMS) seems similar to an alpha numeric paging system. SMS, however, has many advantages over today's paging systems. Paging systems are often limited to very specific regions, while GSM users can receive messages of up to 160 characters in length in more than 100 countries around the world. If a pager is turned off, or has a dead battery it cannot receive messages. If for any reason a message cannot be delivered to a GSM user, the system holds the message until the next time the user inserts their SIM card into a GSM handset. As soon as the handset is turned on, the system delivers any waiting messages.

Short Message Service offers message delivery that is guaranteed to reach the user. If the user's telephone is not turned on the message is held for later delivery. Each time a message is delivered, the network expects to receive an acknowledgment that the message was properly delivered to the handset. If no acknowledgment is received, the system assumes that there was a problem and either re-sends the message or stores it for later delivery.

Mobile Terminated Short Messages

Messages that are sent to a specific GSM user are called mobile terminated short messages. These messages are stored directly on the SIM card. If you borrow a friends handset and receive messages while you are using it, the messages are saved on your SIM, not in your friends handset. This allows you to insert your SIM into a different handset later and still be able to recall the messages you stored previously.

There is a finite amount of storage available on the SIM card for messages. Short messages stay on the SIM card until the user actively deletes them. It is not uncommon for the memory of the SIM to become filled, thus not allowing the delivery of additional messages from the network. When this happens, 'Memory Full' is displayed on the handset. When space is made available on the SIM by deleting messages, the system delivers any waiting messages. Mobile terminated messaged can come from many sources:

- Voicemail message waiting indicators - a message to advise the user that there is a voice message waiting to be picked up from their voicemail box. Voicemail waiting indicators update themselves, a message on the display that says "You have 1 new voice message waiting" would be replaced by a message that says "You have 2 voice messages waiting". In this manner the voicemail message waiting indicator only takes up one space on the SIM card.

- Numeric Pages - many voicemail systems allow the caller to key in a telephone number where they can be reached. This works much like a numeric paging system. The message received might say "Please call (404) 713-2919."

- Live Operator Service Bureau - customers may choose to route their calls to a live operator who takes a message and keys it into a system that is directly linked to the Short Message Service (SMS). A message received from a live operator might say 'Your wife called, she has tickets to the Jimmy Buffet concert tonight. Call her at home to arrange a place to meet before the show' (131 characters).

- Office Based Message Systems - software for the office PC allows a secretary to type messages to an individual or group and send these messages directly to GSM phone users via a desktop computer with modem access into the short message service.

- From The Network Operator - network operators can send messages directly to their customers. This message may be anything from a late payment notice, to an announcement about a new service or feature.

- Other GSM Users - it is possible for a GSM user to send a message directly from their handset to another GSM user.

- Over The Internet - many popular internet sites allow you to send short messages directly to a user over the world wide web.

- Information Service Providers - users can subscribe to information services such as news or stock quotes which are delivered in the form of short messages.

Most GSM handsets will allow telephone numbers included in GSM Short Messages to be dialed automatically. If a message on the display of your phone has a phone number in it, press the "SEND" button to call that number.

Mobile Originated Short Message

GSM short messaging allows users to send messages directly from their handset to another user. This is called mobile originated short message service. Messages can either be keyed into the handset directly, or 'canned' messages can be recalled from memory and sent. It is also possible to send messages to other users via the GSM phone from a laptop computer connected to the handset.

 In order for a message to be routed from mobile handset to the correct Short Message Service Center (SMSC), there must be a service center number programmed into the SIM card. This number can either be placed into the card at the time that the card is programmed by the manufacturer, or it can be programmed into the card via the handset. Since the electrical profile for the card is often developed for the SIM cards prior to the time that the SMSC number is known, customers must often program this code into the handset before they are able to send short messages from their handset.

Executable Short Message

The short message service can also be used to send messages from the network directly to the computer chip that is resident in the SIM card. During the section on activation, for instance, we said that information on the SIM such as the MSISDN and wake-up message were updated over the air. This update is done by sending an Executable Short Message to the SIM card. Executable, in this case, refers to the fact that the message contains a command which must be executed by the SIM card. Almost any memory field on the SIM card can be modified using an executable short message, one notable exception is the Ki. Engineers are developing new ways to use executable short messages for even more than the ability to update the SIM card during activation.

Cell Broadcast

Like a standard short message, a cell broadcast message is an alpha numeric message that can be received on a GSM mobile handset. Cell broadcast works differently than standard short messaging. A cell broadcast message is not sent to an individual, or specific group of individuals; it is sent to anyone who is using their service within the coverage area of the tower (or towers) that transmits the message. A cell broadcast message is not held for later delivery. If your handset is powered up and in the coverage area you get the message, if it is not, you don't get the message.

As an example, imagine a group of three towers along Daytona Beach Florida. Users on the beach might receive a cell broadcast message that says:

"Tidal wave to hit beach in 4 minutes, GET OUT NOW!"

There would be no reason to delay delivery of this message and send it later, the people not in the coverage area of those three towers do not need to receive this message.

Cell broadcast messages can be used to deliver any type of geographically specific message such as emergency warnings, traffic updates or even restaurant advertisements. Unless special arrangements are made, a cell broadcast message can only be initiated by the network operator. Cell broadcast messages are only transmitted on a small group of radio channels. Each message contains a key word identifier, which could be used to designate which messages were received by a user and which ones were ignored. For instance messages with a keyword that identifies an emergency would be received by all users, while only those users who had paid for the service would receive messages with a keyword that indicated sports score updates.

Billing

But When Does The Bill Get Paid?

Chapter 18
Billing

The mobile phone bill is often the only tangible contact between the network operator and the customer. This is especially important since it influences the customer's personal finances. While no one is ever happy to see a bill, an accurate and well formatted bill can significantly lessen any negative reaction.

This chapter is divided into two sections. The first section describes under what conditions a customer is charged. The second section describes the systems used to process and track customer charges and account information.

When Does A Customer Get Charged?

Monthly Service Charges

Generally speaking, a customer can expect to pay a monthly fee for each active account. In addition to this fee, the customer is charged for any extra services or features not included in their monthly service. Since not all customers are billed on the same day, each customer is assigned a specific start and end date. This period is known as the bill cycle. Monthly fees are pro-rated if the service activation does not take place at the beginning of the bill cycle.

Airtime

In most cases customers are charged a per minute fee for each minute spent on an actual telephone call. A chargeable minute is referred to as a minute of "Airtime". Which calls are charged airtime and which are not varies by network operator; in addition some network operators bill by rounding up to the nearest whole minute, while others bill in per second increments.

In North America, unlike other parts of the world, mobile customers are charged for both incoming and outgoing calls. In other words it does not matter who placed the call, the **Mobile Party Pays**. In many other parts of the world, callers are charged a premium for dialing a mobile telephone number from any type of phone. Incoming calls are received on the mobile phone at no charge. In other words, in this environment the **Calling Party Pays**. As previously stated, North America is a Mobile Party Pays environment, customers pay for incoming and outgoing calls.

Customers are not usually charged for call attempts which are not answered or that result in busy signals or recorded network announcements. Note that a call which is answered by an answering machine is charged, as the call was, in fact, answered. Attempts to call a customer who does not answer are not charged;

most network operators do, however, charge for a call that is routed to another number via call forwarding.

Some operators offer the first minute of all incoming calls for free. Some operators package minutes of no charge airtime together with monthly service. A few even offer unlimited calling for a higher monthly service fee.

Toll Charges

In addition to airtime charges, customers are charged for long distance tolls associated with the calls made. The area in which calls are considered "Local" is often different for the mobile telephone network than it is for the landline network.

When a caller is Roaming (using a network other than their home telephone network), the airtime and per minute toll charge may be different than it is in their home market. In most cases roaming customers are not allowed to use "free" airtime from their monthly service package. Since GSM is an international standard, customers may use their service in many parts of North America and indeed many parts of the world. There must, of course be a Roamer Agreement in place between the home network operator and the foreign network operator. For more information see Call Processing in Chapter 11.

In addition to airtime fees, roaming customers are usually charged toll fees for incoming calls delivered to them from their home market. When calling a Texas based customer who is roaming in Ohio, the call is first routed to the Texas MSC by the PSTN and then from the Texas MSC to the Ohio MSC for delivery. The call leg from the Texas MSC to the Ohio MSC is charged as a toll call to the roaming mobile telephone user.

As previously mentioned some services such as call forwarding, call waiting and three party call conference may accrue airtime and toll charges at a higher than normal rate. A customer who places one call on hold to answer another call is paying airtime for both calls.

The Customer Care And Billing System

Although the network and the handsets tend to get most of the attention in any cellular system the Customer Care and Billing System (CCBS) is a huge part of the total package. The CCBS is often referred to as simply "The Billing System", but it actually does much more than just billing. The CCBS encompasses many different areas of functionality. These multiple functions must be packaged and presented in a manner which allows a customer service representative to locate information quickly and efficiently when a customer calls. Some of the functions handled by the CCBS include:

- Customer Account Tracking
- Service and Feature Assignment
- Airtime Rates and Service Pricing
- Inventory Tracking
- Call Detail Rating
- Roamer Data Processing
- Bill Calculation and Formatting
- Collection and Payment Posting
- Late Payments and Notices

Customer Account Tracking

You have already read a lot about customer account setup in chapter 10. In short, the customer's details (name, address, social etc.) are keyed into the CCBS and a new account is set up. A credit check is run, and the customer is assigned a credit class based upon the result. Many network operators ask the customer to choose a password when the account it established. This password is used to authenticate the customer on future calls.

Service and Feature Assignment

The services and features are explained to the customer and the ones they request are included in the account set up. Some network operators have established sophisticated needs assessment programs which allow the customer service representative to pro-actively suggest services or features that match the customers needs. The credit class assigned to a customer may impact which services may be offered to the customer.

Airtime Rates and Service Pricing

This book has not dealt with pricing in great detail, although it is an extremely important issue. To quote one industry expert:

"One of the biggest dangers that current and new operators face is getting the pricing wrong. Overpricing can damage not just sales but the whole Brand too. Under pricing hits the bottom line and can lead to the nightmare of a price war." Saeed Butt

Saeed Butt is Managing Consultant for The Pricing Practice, based in London.

Charges on a customers account can usually be divided into seven types:

1. Rate plan charges

2. Monthly service fees

3. Price per unit of airtime or messages. Many programs include a certain amount of minutes or messages at no extra charge. May include different rates depending on the time of day or day of the week.

4. Services and features (those not included in the rate plan)

5. Long distance fees (these may be charged on a separate bill by the long distance service provider)

6. Roaming charges - fees for service used outside of the home market

7. Other charges & credits (OCC) - typical OCCs include credits for dropped calls and charges for late payment (OCCs may also include bad check fees or special promotions credits)

Inventory Tracking

Inventory tracking in this situation refers to both the actual handsets and the SIM cards. Each card must be accounted for in the CCBS. The inventory tracking module is also the interface that is used to create the SIM order disk.

Call Detail Rating

As previously stated, the network collects and records the details of every call made to or from the mobile telephone. These details are converted into a format that makes sense to the customer who will see them on the bill. A typical bill includes:

- Telephone number dialed or caller's number

- Which city was called
- The time the call started
- How long the call lasted
- The airtime charge
- Toll charges (if any)

Airtime charges may vary depending on the customers rate plan, whether or not there were packaged minutes to be used and what time of day the call was made (peak or off-peak). Once the calls are rated they are stored until it is time to create the actual mobile phone bill. Stored calls can be viewed by the customer service representative thus allowing them to discuss recent calls with a customer, even before the bill is issued.

Roamer Data Processing

After call details are collected for roaming customers, they are sent back to the home carrier to be presented on the customer's bill. However, before the call detail records can be sent to the home carrier, they must first be rated, this allows the serving carrier to determine how much money is owed by the home carrier. Since the call detail records may vary slightly by MSC type, the details are also put into a standardized record exchange format. In North America, this format is called NAIG TAP II. The roamer call rating and record formatting is completed by the visited network operator and then passed to a special clearinghouse to be validated for format and content. The clearinghouse passes the validated record along to the home network operator to be included on the subscribers bill. (For more information see Call Processing, Chapter 11.) Once a customer's roamer call records are received, they are stored until it is time to create the bill.

Billing Machine
Photo Courtesy of HTC Horizon, Myrtle Beach, SC

Bill Calculation and Formatting

Once a month, all stored details relating to airtime, toll and roamer charges are collected for each customer. These details, along with any monthly charges and OCCs are organized and totaled. New bill totals are automatically debited to the customer account. The information is then structured as it will appear on the bill. This layout is called "Bill Print Format." Tapes or electronic transfer of the customer information in bill print format are sent to printers who print, fold and stuff the bills into envelopes and mail them.

Collection and Payment Posting

The mobile telephone bill payment stub is printed with graphics which allow the payment to be posted quickly. In most cases, customer payments are mailed to a "Lock Box" facility. This is a company which specializes in the receipt and posting of payments. Payments posted at the lock box are automatically credited to the customers account.

Late Payments and Notices

Most CCBS systems allow for automatic updates of customer account information. Customers whose accounts are late are updated to past due status. For other mobile telephone systems the next step is a past due letter, which costs the network operator approximately $1.20. With GSM this expense is not necessary. Late payment notices can be made by simply sending the user a Short Message advising the customer that their account is past due.

If a customer account becomes far enough past due, the class of service can be changed to Hotline. No matter what number the customer dials, their call will be routed directly to the collections department!

Future Enhancements

So When Do I Get My Dick Tracy Watch?

Chapter 19
The PCS Vision

Step into life in the future of PCS:

Early in the morning Dan wakes up and checks his personal communications device (PCD). The display alerts him that messages were left for him during the night after he had turned on call screening. Emergency calls would have been delivered to him, but all others are routed to an enhanced voicemail service.

Dan takes his PCD into the bathroom with him, and using the built in speakerphone he listen to his voicemail while he brushes his teeth. Since the voicemail system collected the calling line identification of Dan's PCD, he was not required to enter a mailbox number or password. Some of the messages are important so he saves them to deal with when he arrives at the office later. Dan uses the built in voice recognition feature of his PCD get rid of unimportant messages by simply saying the word "Delete."

Once out of the shower, Dan is ready to take calls so he turns off the "Do not disturb" feature. All the calls he makes or receives while he is at home are connected via an in home base station, similar to those being used for cordless handsets today. Both Dan and his wife, as well as their two teenage children all have their own PCD, each with their own telephone number.

After a quick breakfast, Dan is off to the office. He jumps into his energy efficient battery powered car and backs out of the combination docking station and charger. Dan has, of course, brought his PCD with him, and once he is out of range of his home base station, it locks onto the external GSM network. The PCD beeps to alert him to the change, and a display icon lights to let him know that he will now be paying a per minute rate for calls.

Seconds later the PCD rings, and "Donna" (that's Dan's wife), appears on the display. Donna is calling Dan from her PCD to let him know that he forgot his briefcase. She will be driving right past his office on her way to give a lecture at the local college so she can drop it off for him. Dan thanks Donna and hangs up the phone, pleased that Donna called on her PCD before she left home, no sense in them both having to pay a per minute rate.

On the way into the office, Dan plans his day and schedules reminders so that his PCD will beep and display a message a few minutes prior to each meeting. These reminders are on a central system, allowing Dan's secretary to see what he has planned for the day, and to re-arrange his schedule if required.

Dan will be alerted to any changes that his secretary makes by text messages delivered to his PCD.

At the office, Dan uses the same smart card that is in his PCD to open the gate for the parking deck. He parks his car in a charging dock and heads inside. Dan's PCD is now locked onto the office system. As each change of location is made, the landline telephone network is automatically updated. In this manner, a call to Dan's number will be routed to him where ever he is.

The first order of business for the day is to review and sign off on his office long distance charges. Any long distance calls made from Dan's PCD while he is at the office are automatically billed to his corporate account. He can of course override this if he wants to make a personal long distance call. The same holds true for business calls made from home. Dan also has a method for tracking incoming business calls, this is especially important for calls made when he is in the car paying airtime rates, but is also handy as a reference of calls he made. Dan has established a list of the telephone numbers of all his business contacts, and the list is sorted into three priority levels. Each level has its own distinctive ringer programmed in his PCD, so Dan knows just by listening to the phone ring, if the call is business or personal, and if it is business, which list of clients that call is coming from.

After he finishes his work day, Dan heads out to meet Donna for dinner. Since most of the better restaurants today have built in micro cells like the ones at his office, Dan knows that if Donna is running late, she can reach him at the restaurant.

During dinner, both Dan and Donna activate the Do Not Disturb mode. Somewhere in between the appetizers and the main course, both his and Donna's phone emit a soft chirp. That's their daughter Susan checking in, they smile at each other knowing that in just a few minutes their phones will beep again with a slightly different tone, which will be their son, Art. Art is never quite as punctual as Susan, but both kids are pretty good at remembering to send a message to their folks to let them know that they are OK when away from home. It's not a very hard thing for either of them to do, the massage is stored into the handset and the distribution list for the message contains both Dan and Donna's personal phone numbers. Upon receiving a message from either one of them, Dan's PCD also receives that calling line ID from the kids personal numbers, thus allowing his handset to give a distinctive alert for both.

During the ride home, both Dan and Donna put their phones back into normal mode should anyone need to reach them. Donna just hopes that Dan will remember to put his phone back on Do Not Disturb tonight before he goes to sleep. Many of Dan's customers live in Australia, and the time difference means that their calls come in around 4:00AM. They know they can leave messages about the day's business on Dan's personal number, and that he will get them, even if it is not until hours later.

Chapter 20
Other Enhancements

Creative minds from all over the world work to make sure that GSM takes advantage of all the latest and greatest enhancements possible. GSM has been released in phases, the first release in 1991 was Phase 1. The current release is Phase 2 with Phase 2 Plus soon to be released. Phase 2 Plus includes such enhancements as USSD, the SIM Tool Kit and CAMEL:

- USSD or Unstructured Supplementary Service Data, allows free form messages to be sent from the handset to various parts of the network. In the past, all network messages sent from the handset (except for mobile originated short messages) were of a very specific format. USSD will allow flexibility for the development of new services and features.

- SIM Tool Kit is a collection of enhancements to SIM functionality. These enhancements include allowing the SIM card to pro-actively send information to the handset. In the past the SIM has always been a slave to the handset and could only send information as a response to a query from the handset.

- CAMEL (Customized Applications for Mobile network Enhanced Logic) is a series of enhancements that will allow special features and services to be built into GSM networks on a "services creation node". Camel will afford GSM the same flexibility to build local services "on the fly" that is currently available to the landline network using AIN (Advanced Intelligent Network) platforms.

Many projects that are "in the works" will take advantage of the increased functionality of Phase 2 Plus. Special projects include:

- Dual mode phones - phones that will work both on GSM and another network type. A dual mode AMPS and GSM phone, for instance, would help customers who wanted the advanced capabilities of GSM and the nationwide footprint of AMPS.

- Multi-function SIM cards - the most promising enhancements to GSM functionality is in the area of SIM cards. Soon you will be able to use just one highly secure card to use your handset, get cash from the bank, get into the parking deck at the office, pay for a meal or pick up plane tickets without standing in line.

Enhancements may also include a registry of health related information on the card to alert ambulance drivers or hospitals of information such as allergies or the name of your insurance carrier.

- Pre-paid Smart cards - for those who do not wish to pay for service on a monthly basis, or whose credit does allow the establishment of a regular account prepaid smart cards may be a perfect solution. These cards could be purchased at any convenient store, and eliminate the hassle of payment for the customer and collection for the network operator.

- Calling Name - soon you will be able to see not only the telephone number of the person calling you, but their name as well. Calling name on GSM will work much like it does on a landline phone.

- Enhanced Emergency Services - by triangulating your position between three or more towers, GSM systems will be able to provide very specific information about where an emergency call was made from. This will assist emergency services personnel in getting help to an injured caller faster.

- Home Base Station - imagine a phone that is your cordless when you are at home and your GSM mobile when you are away from home thus allowing you to always receive calls on your handset, but at the lowest possible rate. This feature is closer than you might think!

By planning ahead and working together, the **GSM** community has been able to insure that the future for **G**lobal **S**ystems For **M**obile Communications is Clear, Clean and Bright.

Appendixes

Glossary

GSM Network Operators (North America)

GSM Handset Manufacturers

SIM Card Manufacturers

Distributors of Handsets and Accessories

North American Frequencies

A & B Block Licences

World GSM Committee Members

MoU Members

Glossary

A

A3 Algorithm - used by a GSM network to authenticate a user on the network; the A3 algorithm is the math problem run by both the SIM and the AUC. If the answers that result in both locations match, the customer is approved. If they do not match, the customer is denied access.

A5 Algorithm - used to encrypt and decrypt speech and data that is sent over the air in a GSM network.

A8 Algorithm - stored in the SIM card and at the AUC. The A8 algorithm is used to create another key, the Kc, which in turn is used by the A8 algorithm for voice encryption and decryption.

ACK - (Acknowledgment Message) When a SIM card has successfully received a short message it sends an acknowledgment message back to the network via the handset.

AIN - (Advanced Intelligent Network) A external system used to provide enhanced features and services in some GSM networks.

ALS - (Alternate Line Service) Allows a subscriber to have two voice numbers on the same SIM card.

AM - (Amplitude Modulation) - an analog method used to transmit information over a electromagnetic wave by varying the height of the wave.

AMPS - Advanced Mobile Phone Service - the cell based mobile telephone system that has been used in the United States since 1983. AMPS uses frequency modulation (analog) to transmit un-encrypted voice and speech over the air.

Analog - a method of sending information over radio waves in which the information send is analogous (the same as) the original sound being transmitted. Analog transmission methods have been in use in the United States for more than 50 years, however, the industry began switching to Digital Transmission in 1991.

AUC - (Authentication Center) a sub-database of the HLR where the IMSI/Ki combination is stored. Security algorithms for authentication and creation of the Kc are run in the AUC.

Authentication - the process used to minimize fraudulent or unauthorized use of a GSM network.

B

Bits - individual binary digits used to carry information in a digital network.

BSC - (Base Station Controller) contains all the logic used to control the operations of the BTS, and acts as an interface between the BTS and the MSC.

BSCS - (Business Support and Control System) BSCS is both a generic term used to describe the billing system of a GSM network as well as a brand name of billing software sold by GSI Danet and LHS.

BSS - (Base Station Subsystem) a grouping of network parts that includes the BSC, the BTS (with tower).

BTA - (Basic Trading Area) the smaller of the geographic regions into which PCS licenses in the C,D,E & F blocks were auctioned by the FCC.

BTS - (Base Transceiver Station) the digital radios, antennas and tower used to send and receive information over the common air interface.

C

Call Blocking - limitations placed upon the types of called that can be made; block incoming calls, block outbound calls, block international calls. (See also call restrictions)

Call Forwarding - used in two ways: Unconditional - forwards all calls. Conditional - forwards calls to another number when called number is either busy, unreachable or not answered.

Call Restrictions - limitations placed on a users service, such as no long distance, no international or no incoming calls.

Call Waiting - the ability to put an existing call on hold and accept another incoming call. Users are notified by a beep that a call is waiting. Most networks offer calling line ID on the waiting call. Some services offer the ability to be notified of a third incoming call.

Calling Party Pays - environment in which the person who initiates the call pays the premium for mobile service, even if they are a landline phone calling a mobile. (see also Mobile Party Pays)

CDMA - Code Division Multiple Access - a method for transmitting information over the air in a digital format. Code division multiple access allocated a specific code to each packet of user information, thus allowing more than one user to access the radio channel at the same time. (See also TDMA, IS-95 & JS-008)

Cell - the basic geographic unit of a cellular system. Cells may vary in size depending on terrain, capacity demands etc. Each cell is equipped with a transmitter receiver device for communicating with mobile telephones.

Cell Broadcast - allows network operators to send a message to users who have their phones turned on in a particular cell or group of cells.

Cell Site - the transmitter receiver device used to communicate with mobile telephones in a cellular network. (See also BTS)

Cellular Communication - mobile communications using the principle of cellular; namely division of coverage area into cells, channel re-use and computer assisted hand-offs.

Cell Splitting - a method of increasing capacity or improving signal coverage in a cellular network by dividing an existing cell into two or more cells.

CEPT - Conference of European Postal and Telecommunication Administration - an organization of government run telephone companies in Europe.

Channel Re-use - the ability to use the same radio channel in more than one non-adjacent cell in a mobile telephone network. Channel re-use increases the number of potential channels available on a cellular network.

Cloning - the illegal process of reproducing the ESN and mobile number of a subscribers phone and putting them into a different phone which can then be used for fraudulent calls.

Coding - the process within digital transmission which converts a sample of voice into binary information for transmission

Common Air Interface - the space between the antenna of a cell site and the antenna of a mobile telephone. The air interface can be transmitted using either analog or digital formats.

Coverage Area - the area in which a mobile telephone user can make and receive phone calls on the mobile network.

D

DCS-1800 - a variation on the original GSM standard, differences include power output and the operating frequency of the network.

DCS-1800 handsets cannot be used on a GSM network (and vice versa). Customers may use their service by inserting their SIM card into a compatible phone. Deployed in 10 markets worldwide.

DCS-1900 - a variation on the original GSM standard, differences include power output and the operating frequency of the network. While DCS-1900 handsets cannot be used on a GSM network (and vice versa) customers may use their service by inserting their SIM card into a compatible phone. DCS-1900 is used throughout the United States and Canada.

Dead Spot - also known as a hole; a location in the coverage area where the mobile telephone network does not provide sufficient coverage

Digital - a new method for transmitting information over the air in which information is converted into a series of numbers (digits). Since it is more flexible and can carry more information, digital transmission allow network operators to design systems that are cleaner, quieter, more secure and have even more services and features than previous analog networks.

Duplex - two way communications that allows callers to talk and listen at the same time.

E

EIR - (Equipment Identity Register) a database used to store the IMEIs of locally issued telephones in three files; white (known good), black (known bad or stolen), gray (under test).

Electromagnetic Spectrum - the range of waves that is usually defined in terms of frequencies. The electromagnetic spectrum is divided into functional sections including; radio, ultra violet light, visible light, infra red light, X-rays, cosmic rays, and gamma rays.

Encryption - the process of "scrambling" information using secret keys and mathematical algorithms.

ESMR - Enhanced Specialized Mobile Radio. One of the many technologies being implemented in North America.

ETSI - European Telecommunications Standards Institute - a group that, along with others (CCITT & CEPT), was instrumental in the development of GSM.

F

FCC - Federal Communications Commission, the government body that regulates telecommunications in the United States.

FM - Frequency Modulation - a method of encoding information into radio waves. FM is clearer and quieter than AM, but not as robust as digital.

Frame - a subsection of a TDMA radio channel. Frames are made up of groups of time slots. The number of slots per frame is determined by the network architecture.

Frequency - The number of complete electromagnetic wave cycles that pass a given point in time in one second. Frequencies are measured in terms of Hertz (after Heinrich Hertz, first man to artificially manufacture a radio wave).

G

GSM - (Global Systems for Mobile communications) previously Groupe Speciale Mobile): the mobile telephone system originally designed to offer consistent digital cellular service throughout Europe. GSM was accepted by most countries of the world as the standard for digital communications and is the foundation for DCS-1800 and DCS-1900.

GSM NA - (GSM North America) a group of individuals and companies that are involved with GSM in North America. The group works together to insure uniformity and state of the art in North American GSM systems.

H

Hand-off /Handover - the automatic switching of a call from one radio tower to another. Hand-offs/ handovers insure the strongest possible signal strength as a mobile telephone user travels throughout the coverage area.

Hertz - a unit of measure. The number of complete radio cycles that occur in one second. (see also frequency).

HLR - (Home Location Register) the central database of a GSM network. The HLR stores information required to allow a customer access and support service on the network. A network may have one or more HLR.

Holes - see dead spots

I

IMEI - (International Mobile Equipment Identity) a unique number used to identify GSM/DCS compatible handsets. (see also EIR)

IMSI (International Mobile Subscriber Identity Number) the primary number used to identify a user on the GSM network. The IMSI is used in conjunction with the mobile telephone network for call routing.

IMTS - (Improved Mobile Telephone Service) one of the pre-cellular mobile phone networks used in the United States.

IS-136 - a digital radio format currently being deployed in North America which uses TDMA as it's transport mechanism. (see also TDMA)

IS-95 - a digital radio format created by Qualcomm, currently being deployed in North America. IS-95 uses CDMA as it's transport mechanism in the 800 MHz range. To date this technology has had limited use in real life situations. (see also JS-008)

ISDN - (Integrated Services Digital Network) the new standard for digital communication signal over landline networks.

IWF - (Inter-Working Function) a cabinet full of modems used to convert fax and data information into the appropriate format to be sent to or from a GSM mobile telephone.

J

JS-008 - a digital radio format created by Qualcomm, currently being deployed in North America. JS-008 uses CDMA as it's transport mechanism in the 1900 MHz range.

K

Kc - the number resulting from the Ki and a random number being run through the A-8 algorithm. The Kc is used by the network and the handset and is one of the input variable in the encryption process.

Ki - the secret code used in authentication and encryption of GSM mobile telephone services. Each GSM user is given a unique Ki.

M

MOSMS - (Mobile Originated Short Message Service) a service which provides users the ability to send short messages to another user directly from the handset.

Mobile Party Pays - environment in which any charges are applied to the mobile users' bill. Mobile Party Pays is the standard in North America.(see also Calling Party Pays)

MoU - (Memorandum of Understanding) a legal document that once signed commits the signatory to the standards and specifications for building a GSM network. The first MoU was signed by the members of the GSM committee in 1987.

MS - (Mobile Station) the combination of a SIM card and a handset compatible for use on a GSM/DCS network.

MSC - (Mobile Switching Center) the central switch of a GSM/DCS network. The MSC has three main duties - switch calls, collect call detail records and supervise system operations.

MSISDN -(Mobile Subscriber ISDN) the mobile telephone number used by GSM/DCS networks. In the United States the MSISDN follows the North American Numbering Plan

MTS - (Mobile Telephone Service) the first public mobile telephone network implemented in the United States.

MTSO - the central switch of an AMPS network. The MTSO has many more responsibilities than the MSC in a GSM/DCS network including authentication, call supervision, and hand-off control.

MWI - Message Waiting Indicator - used to alert customers that a message has been left in their voice mail box. In GSM networks the MWI is delivered via the short message channel.

N

NACK - (Negative Acknowledgment) sent by a GSM handset to the network when the SIM card was unable to process received short message.

NMC - (Network Maintenance Center) facility that allows 24 hour monitoring, as well as test and maintenance routines for all parts of a GSM network. (See also OMC-R & OMC-S)

North American Numbering Plan - A standard format for telephone numbers used throughout the United States. (NPA) NXX-line #

NSS - (Network Subsystem) - a group of network components including the MSC, HLR, VLR and EIR.

O

Off Peak - the period of time after the business day during which many carriers discount long distance or airtime charges.

OMC-R - (Operations and Maintenance Center) provides sophisticated alarms and system monitoring tools which can be used by network engineers to constantly monitor all functions of the radio side of a GSM/DCS network.

OMC-S - (Operations and Maintenance Center) provides sophisticated alarms and system monitoring tools which can be used by network engineers to constantly monitor all functions of a GSM/DCS MSC.

One-way - communication that travels in one direction only. For example a radio station broadcast. (see also two way communication)

P

Packet - a collection of data which is transmitted over a digital network in a 'burst'.

PCMCIA - (Personal Computer Memory Card Industry Association) A credit card sized modem used with many portable computers and faxes.

PCN - (Personal Communications Networks) refers to communications licenses issued in the United Kingdom.

PCS - (Personal Communications Service) a collection of services and capabilities which provides users flexibility of access, and personal mobility. PCS can be offered by a combination of wireless and wireline networks. The basic concept is a telephone number for a person, not a location.

PCS Data Card - a special card which plugs into the PCMCIA type II data slot on a computer or fax machine and connects to a GSM phone. This card enables users to send data and faxes via their mobile phone.

Peak - the part of the business day/week when customers can expect to pay full rates for long distance and mobile telephone services.

PIN - (Personal Identification Number) Customers may set a PIN number to insure against unauthorized use of their SIM card. If the PIN number is keyed incorrectly three times in a row the card becomes blocked. A higher access level code is required to unblock the phone (see also PUK).

POTS - the Plain Old Telephone System (see also PSTN)

PSTN - (Public Switched Telephone Network) The landline telephone system.

PUK - (Personal Unblocking Key) used to reset the PIN number on a SIM card if it has been locked. Keying the PUK code incorrectly ten times in a row will result in the card being permanently disabled. (See also PIN)

R

Radio Tower - holds the radio transmitter and receiver antennas in a mobile telephone network.

Roam - the ability to use a mobile telephone outside of the area considered to be the users home service area. In an automatic roaming scenario, all call charges appear on the bill received from the home mobile telephone carrier.

RWC - the Repair and Warranty Center. Used By BellSouth Mobility DCS for handset returns/ repairs.

S

Sampling - in order to limit the amount of information sent over the GSM network, smaller samples of speech are taken and converted into packets of digital information. The device that samples and rebuilds speech in a GSM/DCS network is known as a voice encoder/decoder (vo-coder)

Short Message - an alphanumeric message of up to 160 characters that can be sent to a GSM/DCS subscriber. Messages are delivered directly to the SIM card and are read on the display of the handset.

SIM -(Subscriber Identity Module) the 'smartcard' which contains a users service information. The SIM card has four main jobs in the GSM network. 1. authentication 2. storage of data 3. assist in encryption process 4. subscriber protection via PIN/PUK.

Simplex - a form of two way communications, in which a single communications path is shared by both parties in the conversation, they must take turns talking.

Sine Wave - a graphic representation of the electromagnetic wave form. (see also frequency)

Smart Card - a credit card sized piece of plastic with a microprocessor built into it. SIM cards are one type of smart card, but not all smart cards are SIM cards.

SMS - (short message service) allows the transmission and receipt of short messages. (see also SMSC)

SMSC - the facility used to house and deliver short messages (see also SMS)

Standby Time - the time when the mobile telephone is turned on and attached to the network, but does not have a call in progress. (see also talk time)

Switch - The central control computer of a mobile telephone network. (See also MSC and MTSO)

T

Talk Time -The time when a mobile telephone call is in progress.

Talk time uses considerably more battery power than does standby time (see also standby time)

TDMA - Time Division Multiple Access - a method for transmitting information over the air in a digital format. Time division multiple access divides a radio channel into time slots and allocates customer information to these slots, thus allowing more than one user to access the radio channel at the same time. IS-136, ESMR and GSM all use TDMA as their radio transport mechanism. (See also CDMA)

Three Party Call Conference - a value added service which allows multiple parties to be connected on a single "conference' call.

TIMSI - The TIMSI is used as an alias for the IMSI. In this way GSM do not have the security risk of sending the actual IMSI over the air. The network automatically changes the TIMSI from time to time.

Transceiver- a radio transmitter receiver device.

Trunking- a system of channel allocation that allows many users to separate on the same group of radio channels. Users take turns using the next available channel.

Time Slot- a space in time used for transmitting digital information over the air using TDMA

Trunks- groups of cables or fiber optic circuits used to carry telecommunications signals from one location to another.

Two-way- communication that travels in two directions Two way communication can either be simplex or duplex (see also simplex, duplex and one way communication)

V

Vo-coder - The voice encoder decoder device used to sample the voice sent over a digital mobile telephone network.

VLR - (Visitor Location Register) the database that holds information about users who are currently registered in a service area. Information in the VLR is updated once a day.

VMS - Voice Mail Service - service based upon a system that is capable of playing a greeting and record messages which are placed in a virtual mailbox. The messages can later be retrieved telephonically by the owner of the mailbox. Users are alerted that a message has been placed in their mailbox via a message waiting indication (MWI). In GSM/DCS networks the MWI is delivered via SMS.

WHO'S WHO LIST

GSM Network Operators

Aerial - http://www.teldta.com/tds/portable/

Tampa/Orlando Fl., Houston Tx, Columbus Oh., Minneapolis Mn., Kansas City

8410 West Bryn Mawr Avenue
Suite 1100
Chicago, IL 60631
Phone (773) 864-4000
Fax (773) 864-4397

Airadigm Communications (Einstien) -
http://www.airadigm.com

Wisconsin

2301 Kelbe Drive
Little Chute, WI 54140
Phone 1-800-745-1818

American Personal Communications -
http://www.sprintspectrum-apc.com/

Washington, DC, Baltimore

1208 18th St.
Washington, DC 20036 (Retail Store)
Phone 1-800-311-4220

BellSouth Mobility DCS - http://www.bellsouthdcs.com

North Carolina, South Carolina, East Tennessee, parts of Georgia

3353 Peachtree Road NE
Suite 300
Atlanta, GA. 30326
Phone (404) 841-2000
Fax (404) 841-2045

DigiPH Communication - http://www.digiph.com/

Microcell Solutions, Inc. - http://www.fido.ca/

Montreal, Quebec City, Ottawa, Toronto, Vancouver

1250 Rene-Levesque Blvd West
Suite 400
Montreal Canada H3B4W8
Phone (514) 933-FIDO (3436)
Fax - (514) 925- 7205

Omnipoint Communications, Inc.- http://www.omnipoint.com/

New York (metropolitan area)

16 Wing Drive
Cedar Knolls, NJ 07927
1-888-BUY-OMNI
(1-888-289-6664)

PacBell Mobile Systems -http://pacbell.mobile.com

Most major cities in California, Las Vegas

4420 Rosewood Drive, Building 2, 3rd Floor
Pleasanton, CA 94588
Phone (510) 227-3000
Fax (510) 227- 7412

Pocket - http://www.pocketcomm.com/

Chicago, Detroit, Little Rock, Omaha, New Orleans,
St. Louis, Dallas, Las Vegas, Honolulu, Guam

2550 M. Street N.W. Suite 200
Washington, DC 20037
Phone: (202) 496-4300
Fax: (202) 331-0809

Powertel- http://www.icel.com/POWERTEL/

Jacksonville, Atlanta, Memphis, Birmingham, parts of Kentucky

1233 O.G. Skinner Drive
West Point GA 31833
Phone (706) 645-9967
Fax (706) 645-9523

VoiceStream - http://www.wwireless.com

Iowa, Oregon, Oklahoma, Idaho, New Mexico, Colorado, Utah, Wyoming, South Dakota, North Dakota

2000 NW Sammamish Rd. #100
Issaquah, WA 98027
Phone 1-888-873-7326
Fax (206) 313-7898

Handset Manufacturers

Ericsson - http://www.ericsson.nl

740 E. Campbell Road
Richardson Tx 75081
Phone; (972) 437-8188
Fax (972) 705-7888

Motorola - http://www.mot.com/

One Continental Towers
1701 Golf Road
Rolling Meadows, Il 60008
Phone (800) 291-4685
Fax (602) 441-2712

Mitsubishi - http://www.mitsubishi.com

3805 Crestwood Pkwy, Suite 350
duluth, GA 30096
Phone (770) 638-2100

Nokia - http://www.nokia.com

2300 Valley View Lane
Suite 100
Irving, Tx 75062
Phone (972) 257-9800
Fax (972) 257-9988

Nortel - http://www.nortel.com/wireless

2221 Lakeside Blvd
Richardson, TX 75082
Phone (972) 684-1000
Fax (972) 684-3942

Sagem -

4, rue du Petit Albi
95800 Cergy Saint Cristopphe
B.P. 8448 - 95807 Cergy Pontoise Cedex France

Siemens - http://www.siemens-wireless.com

2220 Campbell Creek Blvd
Richardson, TX
Phone (972) 997-7300

SIM Card Manufacturers

Gemplus Corp. - http://www.gemplus.com

6600 LBJ Freeway
Suite 109
Dallas Tx 75240
Phone (972) 726-1840
Fax (972) 726- 1868

Giesecke & Devrient America, Inc. (G&D)

11419 Sunsuet Hills Road
Reston Virginia, 20190
Phone (703) 709-5828
Fax (703) 810-8459

Orga Card Systems- http://www.orga.com

Station Square Two, Suit 107
Paolia Pa 19301
Phone (610) 993-9810
Fax (610) 993-8641

Sclumberger - http://www.slb.com/

8360 E. Via De Ventura
Building L-200
Scottsdale, AZ 85258
Phone (602) 905-5500
Fax (602) 905-5536

Distribution & Accessories

Brightpoint - http://www.brightpoint.com

6402 Corporate Drive
Indianapolis, IN 46278
Phone (800) 297-6100
Fax (317) 290-9642

Cell Star

1730 Briercroft Court
Carrollton, TX 75006
Phone (972) 466-5377

Ora Electronics - http://www.orausa.com

9410 Owensmouth Avenue
Chatsworth, CA 91311
Phone (818) 772-2700
Fax (818) 718- 8626

Tessco - http://www.tessco.com

34 Loveton Circle
P.O. Box 5100
Sparks, MD 21152-5100
Phone (800) 472-7373
Fax (410) 472-7575

North American PCS Frequencies

Transmit Side

Block A	MTA	1850 MHz to 1865 MHz
Block B	MTA	1870 MHz to 1885 MHz
Block C	BTA	1895 MHz to 1910 MHz
Block D	BTA	1865 MHz to 1870 MHz
Block E	BTA	1885 MHz to 1890 MHz
Block F	BTA	1890 MHz to 1895 MHz

Receive Side

Block A	MTA	1930 MHz to 1945 MHz
Block B	MTA	1950 MHz to 1965 MHz
Block C	BTA	1975 MHz to 1990 MHz
BlocK D	BTA	1945MHz to 1950 MHz
Block E	BTA	1965 MHz to 1970 MHz
Block F	BTA	1970 MHz to 1975 MHz

A& B Block Licenses

Pop	(M)#	Block	Market	High Bidder	$/pop
	1	A	New York	Pioneer's Preference*	
26.41	1	B	New York	WirelessCo	16.76
	2	A	Los Angeles	Pioneer's Preference*	
19.15	2	B	Los Angeles	PacTel	25.78
12.07	3	A	Chicago	AT&T	30.88
12.07	3	B	Chicago	PrimeCo	31.90
11.89	4	A	San Francisco	WirelessCo	17.37
11.89	4	B	San Francisco	PacTel	17.00
10.00	5	A	Detroit	AT&T	8.12
10.00	5	B	Detroit	WirelessCo	8.61
9.75	6	A	Charlotte	AT&T	6.83
9.75	6	B	Charlotte	BellSouth	7.27
9.69	7	A	Dallas	PrimeCo	9.03
9.69	7	B	Dallas	WirelessCo	9.12
9.45	8	A	Boston	AT&T	12.87
9.45	8	B	Boston	WirelessCo	13.44
8.93	9	A	Philadelphia	AT&T	9.07
8.93	9	B	Philadelphia	PhillieCo	9.52
	10	A	Washington	Pioneer's Preference*	
7.78	10	B	Washington	AT&T	27.23
6.94	11	A	Atlanta	AT&T	28.58
6.94	11	B	Atlanta	GTE	26.60
5.99	12	A	Minneapolis	WirelessCo	6.63
5.99	12	B	Minneapolis	Aerial	6.11
5.42	13	A	Tampa	Aerial	16.57
5.42	13	B	Tampa	PrimeCo	18.33
5.19	14	A	Houston	Aerial	16.16
5.19	14	B	Houston	PrimeCo	15.93

Pop	(M)#	Block	Market	High Bidder	$pop
5.14	15*	A	Miami	WirelessCo	25.64
5.14	15	B	Miami	PrimeCo	24.53
4.95	16	A	Cleveland	Ameritech	17.59
4.95	16	B	Cleveland	AT&T	17.36
4.93	17	A	New Orleans	WirelessCo	19.07
4.93	17	B	New Orleans	PrimeCo	18.17
4.72	18	A	Cincinnati	AT&T	8.89
4.72	18	B	Cincinnati	GTE	9.06
4.66	19	A	St. Louis	AT&T	25.48
4.66	19	B	St. Louis	WirelessCo	24.51
4.54	20	A	Milwaukee	WirelessCo	18.73
4.54	20	B	Milwaukee	PrimeCo	18.94
4.10	21	A	Pittsburgh	WirelessCo	7.00
4.10	21	B	Pittsburgh	Aerial	7.72
3.88	22	A	Denver	WirelessCo	16.60
3.88	22	B	Denver	GTE	16.62
3.85	23	A	Richmond	AT&T	8.75
3.85	23	B	Richmond	PrimeCo	8.59
3.83	24	A	Seattle	GTE	27.79
3.83	24	B	Seattle	WirelessCo	27.48
3.62	25	A	Puerto Rico	AT&T	15.70
3.62	25	B	Puerto Rico	Centen	15.09
3.56	26	A	Louisville	AT&T	13.85
3.56	26	B	Louisville	WirelessCo	13.10
3.51	27	A	Phoenix	AT&T	22.32
3.51	27	B	Phoenix	WirelessCo	21.54
3.47	28	A	Memphis	Powertel	12.46
3.47	28	B	Memphis	SWBell	12.46
3.24	29	A	Birmingham	WirelessCo	10.97
3.24	29	B	Birmingham	Powertel	10.87

Pop	(M)#	Block	Market	High Bidder	$/pop
3.06	30	A	Portland	Western	11.16
3.06	30	B	Portland	WirelessCo	11.16
3.02	31	A	Indianapolis	WirelessCo	23.34
3.02	31	B	Indianapolis	Ameritech	23.56
3.01	32	A	Des Moines	Western	7.35
3.01	32	B	Des Moines	WirelessCo	7.00
2.99	33	A	San Antonio	WirelessCo	18.21
2.99	33	B	San Antonio	PrimeCo	17.39
2.91	34	A	Kansas City	Sprint	8.11
2.91	34	B	Kansas City	Aerial	8.10
2.78	35	A	Buffalo	WirelessCo	6.80
2.78	35	B	Buffalo	AT&T	7.15
2.57	36	A	Salt Lake City	Western	17.82
2.57	36	B	Salt Lake City	WirelessCo	17.95
2.27	37	A	Jacksonville	Powertel	20.22
2.27	37	B	Jacksonville	PrimeCo	19.56
2.15	38	A	Columbus	AT&T	10.39
2.15	38	B	Columbus	Aerial	10.34
2.11	39	A	El Paso	Western	4.08
2.11	39	B	El Paso	AT&T	4.08
2.05	40	A	Little Rock	SWBell	6.21
2.05	40	B	Little Rock	WirelessCo	6.01
1.88	41	A	Oklahoma	Western	5.92
1.88	41	B	Oklahoma	WirelessCo	7.00
1.86	42	A	Spokane	Poka	3.05
1.86	42	B	Spokane	WirelessCo	3.32
1.77	43	A	Nashville	WirelessCo	9.26
1.77	43	B	Nashville	AT&T	8.95
1.72	44	A	Knoxville	AT&T	6.18
1.72	44	B	Knoxville	BellSouth	6.47

Pop	(M)#	Block	Market	High Bidder	$/pop
1.66	45	A	Omaha	AT&T	2.80
1.66	45	B	Omaha	Cox	3.06
1.12	46	A	Wichita	AT&T	3.91
1.12	46	B	Wichita	WirelessCo	4.36
1.11	47	A	Honolulu	Western	20.18
1.11	47	B	Honolulu	PrimeCo	19.56
1.10	48	A	Tulsa	SWBell	16.02
1.10	48	B	Tulsa	WirelessCo	15.32
.55	49	A	Alaska	Aerial	1.82
.55	49	B	Alaska	GCI	3.00
.18	50	A	Guam	Poka	0.61
.18	50	B	Guam	AmerPort	0.81
.05	51	A	Amer Samoa	SSeas	4.57
.05	51	B	Amer Samoa	ComIntl	4.85

* The A block PCS licenses in New York, Los Angeles, and Washington D.C. were awarded under the FCC's Pioneer's Preference Rule. The New York license was awarded to Omnipoint for $347.5 million, Cox Communications was awarded the Los Angeles license for $251.9 million, and American Personal Communications was awarded the Washington D.C. license for $102.3 million.

World GSM Committee Members

COUNTRY	COMPANY	TECHNOLOGY
MOZAMBIQUE	Telecomunicacoes de Mocanbique	900 / 1800
SWITZERLAND	Swiss Telecom PTT	900 / 1800
ALBANIA	Albanian Mobile Communications	GSM900
ANDORRA	Servei De Tele. D'Andorra	GSM900
ARMENIA,	Rep of ArmenTel	GSM900
AUSTRALIA	Optus Communications	GSM900
AUSTRALIA	Telstra Corporation Ltd	GSM900
AUSTRALIA	Vodafone Pty Ltd	GSM900
AUSTRIA	Max.Mobil Telekommunikations Service	GSM900
AUSTRIA	Mobilkon Austria AG	GSM900
AZERBAIDJAN	Rep.ofAzercell	GSM900
BAHRAIN	Batelco	GSM900
BANGLADESH	GrameenPhone Ltd	GSM900
BELGIUM	Belgacom Mobile	GSM900
BELGIUM	Mobistar-Telemate	GSM900
BOSNIA HERZEGOVINA	Public Enterprise PTT BiH	GSM900
BRUNEI DARUSSALAM	DataStream Technology	GSM900
BULGARIA	MobiTel AD	GSM900
BURKINA FASO	Onatel	GSM900
CAMBODIA,	Kingdom ofCamGSM	GSM900
CAMEROON	Ministry P&T Cameroon	GSM900
CHINA	Peoples Rep ofChina Telecom	GSM900
CHINA	Peoples Rep ofChina Unicom	GSM900
CROATIA	HPT Croatian P&T	GSM900
CYPRUS	Cyprus Telecommunicatins Authority	GSM900
CZECH REPUBLIC	Eurotel Praha Ltd	GSM900
CZECH REPUBLIC	RadioMobil	GSM900

COUNTRY	COMPANY	TECHNOLOGY
DENMARK	Sonofon	GSM900
DENMARK	Tele Danmark Mobil A/S	GSM900
EGYPT	Arento	GSM900
ESTONIA	Estonian Mobil Telephone Co	GSM900
ESTONIA	Radiolinja Estonia Ltd	GSM900
ESTONIA	Ritabell	GSM900
ETHIOPIA	Ethiopia Telecommunications Authority	GSM900
FED REP YUGOSLAVIA	Mobile Telecommunications	GSM900
FED REP YUGOSLAVIA	ProMonte GSM	GSM900
FIJI	Vodafone Fji Ltd	GSM900
FINLAND	Alands Mobitelfon AB	GSM900
FINLANDOy	Radiolinja AB	GSM900
FINLAND	Telecom Finland	GSM900
FRANCE	France Telecom Mobiles	GSM900
FRANCESFR	Coifra	GSM900
FRENCH POLYNESIA	Tikiphone SA	GSM900
FRENCH WEST INDIES	France Caraibe Mobiles	GSM900
FYROM (Macedonia)	PTT'Makedonija'	GSM900
GEOGRIA	Geocell Ltd	GSM900
GEORGIA	Magticom Ltd	GSM900
GEORGIA	Superphone Ltd	GSM900
GERMANY	DeTeMobil	GSM900
GERMANY	Mannesmann Mobilfunk GmbH	GSM900
GHANA	Francis Walker (GH) Ltd	GSM900
GHANA	ScanCom Ltd	GSM900
GIBRALTAR	Gibraltar Telecommunications Int Ltd	GSM900
GREECE	Panafon SA	GSM900
GREECESTET	Hellas Telecommunications	GSM900
GUERNSEY	Guernsey Telecoms	GSM900
GUINEA	International Wireless Guinea Ltd	GSM900

COUNTRY	COMPANY	TECHNOLOGY
HONG KONG	Hongkong Telecom CSL	GSM900
HONG KONG	Hutchison Telephone Ltd	GSM900
HONG KONG	SmarTone Moible Comms Ltd	GSM900
HUNGARY	Pannon GSM Telecommunications	GSM900
HUNGARY	Westel900 Mobil Tavkozlesi Rt	GSM900
ICELAND	Telecom Iceland	GSM900
INDIA	Bharti Cellular Ltd	GSM900
INDIA	Bharti Telenet Ltd	GSM900
INDIA	Birla AT&T Communications Ltd	GSM900
INDIA	BPL Mobile Communicaitons Ltd	GSM900
INDIA	BPL USWEST Cellular	GSM900
INDIA	Cellular Communications India Ltd	GSM900
INDIA	Escotel Mobile Communications Ltd	GSM900
INDIA	Evergrowth Telecom	GSM900
INDIA	Fascal Limited	GSM900
INDIA	Hexacom India Limited	GSM900
INDIA	Hutchison Max Telecom PVT Ltd	GSM900
INDIA	J T Mobiles Ltd	GSM900
INDIA	Koshika Telecom Ltd	GSM900
INDIA	Modi Telstra	GSM900
INDIA	ModiCom Network Private Ltd	GSM900
INDIA	Reliance Telecom Private Ltd	GSM900
INDIA	RPG Celular Services	GSM900
INDIA	SkyCell Communications Ltd	GSM900
INDIA	Sterling Cellular Ltd (SCL)	GSM900
INDIA	TATA Communications Ltd	GSM900
INDIA	Usha Martin Telekom Ltd	GSM900
INDONESIA	Exelcom	GSM900
INDONESIAPT	Satelindo	GSM900
INDONESIAPT	Telekomunikasi Selular	GSM900

COUNTRY	COMPANY	TECHNOLOGY
IRAN	Celcom Iran	GSM900
IRAN	KIFZO	GSM900
IRAN	TCI	GSM900
IRELAND	Eircell	GSM900
IRELAND	Esat Digifone	GSM900
ISLE OF MAN	Manx Telecom	GSM900
ITALY	Omnitel Pronto Italia	GSM900
ITALY	Telecom Italia Mobile	GSM900
IVORY COAST	Comstar Cellular Network	GSM900
IVORY COAST	Loteny Telecom	GSM900
IVORY COAST	Societe Ivorienne De Mobiles (SIM)	GSM900
JERSEY	Jersey Telecommunications	GSM900
JORDAN	JMTS	GSM900
KENYA	Kenya Posts & Telecoms Corp	GSM900
KUWAIT	Mobile Telecommunications Co	GSM900
LA REUNION	Societe Reunionnaise de Radiotelephone	GSM900
LAO	Shinawatra International Ltd	GSM900
LATVIA	Baltcom GSM	GSM900
LATVI	Alatvian Moible Telephone Co	GSM900
LEBANON	FTML	GSM900
LEBANON	LibanCell	GSM900
LESOTHO	Vodacom Lesotho (Pty) Ltd	GSM900
LIBYA ED MADAR	Telecomm Company	GSM900
LITHUANIA	OMNITEL	GSM900
LITHUANIA	UAB Mobilios Telekomunikacijos	GSM900
LUXEMBOURG	P&T Luxembourg	GSM900
MACAU	CTM	GSM900
MADAGASCAR	Sacel Madagascar SA	GSM900
MALAWI	Telekom Network Ltd	GSM900
MALAYSIA	Binariang Communications Sdn	GSM900

COUNTRY	COMPANY	TECHNOLOGY
MALAYSIA	Celcom/Cellular Comms Network	GSM900
MALTA	Telecell Ltd	GSM900
MAURITIUS	Cellplus Mobile Communications Ltd	GSM900
MONACO	Office des Telephones	GSM900
MONGOLIA	MobiCom	GSM900
MOROCCO	ONPT	GSM900
NAMIBIA	MTC	GSM900
NETHERLANDS	Libertel BV	GSM900
NETHERLANDS	PTT Telecom	GSM900
NEW CALEDONIA	Telecom New Caledonia	GSM900
NEW ZEALAND	BellSouth New Zealand Ltd	GSM900
NORWAY	NetCom GSM AS	GSM900
NORWAY	Telenor Mobil	GSM900
OMAN	Sultanate ofMinistry PTT	GSM900
PAKISTAN	Mobilink-PMCL	GSM900
PHILIPPINES	Globe Telecom	GSM900
PHILIPPINES	Isla Communications Co Inc	GSM900
POLANDERA	GSM, Polska Telefonia Cyfrowa	GSM900
POLAND	Polkomtel SA	GSM900
PORTUGAL	Telecel	GSM900
PORTUGAL	TMN	GSM900
QATAR	Q-Tel	GSM900
ROMANIA	MobiFon SA	GSM900
ROMANIA	Mobil Rom	GSM900
RUSSIA	DonTeleCom	GSM900
RUSSIA	Ermak RMS	GSM900
RUSSIA	Extel Mobile Com System	GSM900
RUSSIAKB	Impuls	GSM900
RUSSIA	Mobile TeleSystems (MTS)	GSM900
RUSSIA	North-West GSM	GSM900

COUNTRY	COMPANY	TECHNOLOGY
RUSSIA	United Telecom (US West)	GSM900
RUSSIA	Uratel	GSM900
SAUDI ARABIA	Electronic Application Est - alJawwal	GSM900
SAUDI ARABIA	Ministry of PPT - AlJawwal	GSM900
SENEGAL	Sonatel	GSM900
SEYCHELLES	Cable & Wireless (Seychelles)	GSM900
SINGAPORE	MobileOne (Asia) Pte Ltd	GSM900
SINGAPORE	Singapore Telecom	GSM900
SLOVAK REPUBLIC	EuroTel	GSM900
SLOVAK REPUBLIC	GlobTel GSM	GSM900
SLOVENIA	Mobitel DD	GSM900
SOUTH AFRICA	MTN	GSM900
SOUTH AFRICA	Vodacom Group Pty Ltd	GSM900
SPAIN	Airtel	GSM900
SPAIN	Telefonica Moviles	GSM900
SRI LANKA	MTN Networks (Private) Ltd	GSM900
SUDAN	Sudatel	GSM900
SWEDEN	COMVIQ GSM AB	GSM900
SWEDEN	Europolitan AB	GSM900
SWEDEN	Telia Mobile	GSM900
SYRIA	Syrian Telecommunication Est	GSM900
TAIWAN	Chunghwa Telecom LDM	GSM900
TANZANIA	Tri Telecommunicatins Ltd	GSM900
THAILAND	Advanced Info Service PLC	GSM900
TURKEY	Turk Telecom	GSM900
UGANDA	CelTel	GSM900
UK	Cellnet	GSM900
UK	Vodafone ltd	GSM900
UKRAINE	Ukrainian Mobile Communications	GSM900
UKRAINE	Ukrainian Radio Systems	GSM900

COUNTRY	COMPANY	TECHNOLOGY
United Arab Emirates	Emirates Telecom Corp-ETISALAT	GSM900
UZBEKISTAN	Buztel	GSM900
UZBEKISTAN	Coscom	GSM900
UZBEKISTAN	Uzmacom	GSM900
VIETNAM	VNPT	GSM900
ZAIRE	African Telecomm Network	GSM900
ZIMBABWE	Posts & Telecommunications	GSM900
CANADA	Microcell Telecommunications Inc	GSM1900
USA	American Personal Communications	GSM1900
USA	American Portable Telecommunications	GSM1900
USA	BellSouth Mobility DCS	GSM1900
USA	Omnipoint Corporation	GSM1900
USA	Pacific Bell Mobile Services	GSM1900
USA	PCS One Inc	GSM1900
USA	Pocket Communications	GSM1900
USA	Powertel PCS Partners	GSM1900
USA	Western Wireless Corporation	GSM1900
FINLAND	Finnet Group	GSM1800
FINLAND	Telivo Ltd	GSM1800
FRANCE	Bouygues Telecom	GSM1800
GERMANY	E-Plus Mobilfunk	GSM1800
HONG KONG	Mandarin Communications Ltd	GSM1800
HONG KONG	New World PCS	GSM1800
HONG KONG	P Plus Communications	GSM1800
HONG KONG	Pacific Link	GSM1800
HONG KONG	Peoples Telephone Company Ltd	GSM1800
MALAYSIA	Mutiara Telecommunications Sdn	GSM1800
MALAYSIA	Sapura Digital SDN BHD	GSM1800
MALAYSIA	Telecom Cellular Sdn Bhd	GSM1800
THAILAND	Total Access Communications Co	GSM1800

COUNTRY	COMPANY	TECHNOLOGY
THAILAND	Wireless Communications Service Co	GSM1800
UK	Orange PCS Ltd	GSM1800
UK	One-2-One	GSM1800
UKRAINE	Bancomsvyaz	GSM1800

Telecomunications Administration MoU Members

AUSTRALIA	Austel
FINLAND	Telecomms Admin Centre
FRANCE	Ministere des PTE/DRG
GERMANY	BMPT
HUNGARY	Communications Authority of Hungary
INDIA	TEC
ITALY	Istituto Superiore PT
LEBANON	Ministry of Post & Telecomms
NETHERLANDS	Ministry of Transport HDTP
POLAND	Ministry of Post & Telecomms
PORTUGAL	ICP
RUSSIAMinistry	P&T Russian Federationr
SLOVENIA	Telecommunication Administration
SOUTH AFRICA	Dept of Posts & Telecomms
SPAIN Director	Generale de Telecomm
THAILAND	Telephone Organisation of Thailand
UK	DTI

Numerics

50 MHz Rule 23
911 37, 158

A

A3 algorithm 110
A5 Algorithm 117
A8 Algorithm 117
Activation 99
Advice Of Charge 129
AIG 12
Air Interface 15, 43
Airtime 143
Airtime Rates 146
Alert Tones 55
Algorithm 49
Alkaline 51
ALS 128
AMPS 8, 15, 29, 128
Analog 15
Antenna Types 53
APIG 12
AT&T 29
Attach 119
AUC 35, 38, 41, 95, 110
Auctions 21, 25
Authentication 49, 107, 110
Auto Redial 53

B

Batteries 51
BAUD Rate 131
Bell Labs 8
Bill Formatting 148
Billing 141, 143
Billing System 43, 99, 144
Black List 40
BSC 42
BSS 35, 42
BTA 21, 101
BTS 42

C

D

E

Eavesdropping 17, 117
ECOIG 12
EFR 17
EIG 12
EIR 35, 40, 41
Encoding 16
Encryption 43, 49, 113, 117
Ericsson 59, 61, 63
Error Correction 17
ESMR 25, 28, 29
ETSI 10, 11, 35
Executable Short Message 139

F

Fax & Data 129, 131
FCC 5, 12, 19, 21
Features 102, 121
Forbidden PLMN List 107
Fraud 10, 15
Frequency Blocks 19
Full Size 48

G

Gateway 43
Gray List 40
Gray list 40
Greeting Message 103
GSM 10, 13, 31, 158
GSM MoU 10
GSM MoU Association 11, 12
GSMNA 12

H

Hand Off 8
Hands Free 54
Handset Personalization 88, 98
Handsets 50
Health and Safety 89
Hearing Aids 89
Hertz 4
HLR 35, 36, 37, 38, 41, 43, 95, 102

N

O

P

Telemarketing 98
TMDA 117
Toll 144
Towers 89
Transceiver 4
Transciever 4
Transport Mechanism 25
Two Phones One Number 128

U

USSD 157

V

VLR 35, 38, 39, 41, 108
VLR Record Setup 109
VMS 45
Vo-coder 16, 51
Voicemail 135, 136

W

White List 40
Wireless 3